Dimensional Analysis and
Intelligent Experimentation

Dimensional Analysis and Intelligent Experimentation

Andrew C Palmer

National University of Singapore, Singapore

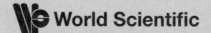

World Scientific

NEW JERSEY · LONDON · SINGAPORE · BEIJING · SHANGHAI · HONG KONG · TAIPEI · CHENNAI

Published by

World Scientific Publishing Co. Pte. Ltd.

5 Toh Tuck Link, Singapore 596224

USA office: 27 Warren Street, Suite 401-402, Hackensack, NJ 07601

UK office: 57 Shelton Street, Covent Garden, London WC2H 9HE

British Library Cataloguing-in-Publication Data
A catalogue record for this book is available from the British Library.

ISBN-13 978-981-270-818-2
ISBN-10 981-270-818-9
ISBN-13 978-981-270-819-9 (pbk)
ISBN-10 981-270-819-7 (pbk)

Typeset by Stallion Press
Email: enquiries@stallionpress.com

Printed by Fuisland Offset Printing (S) Pte Ltd. Singapore

Preface

Dimensional analysis is a clever strategy for extracting knowledge from a remarkably simple idea, nicely stated by Richardson

> "... that phenomena go their way independently of the units whereby we measure them."

Within its limits, it works excellently, and makes possible astonishing economies in effort. The limits are soon reached, and beyond them it cannot help. In that it is like a specialized tool in carpentry or cooking or agriculture, like the water-driven husking mill I saw in Viet Nam a few days before writing this preface, which husks rice elegantly and admirably but cannot do anything else.

Some of the motivation for writing this book came from lectures I gave to first-year students at Cambridge in 1996. I had used dimensional analysis many times before, but when I came to teach it I looked for a textbook. I was disappointed in the existing books, particularly in their treatment of fundamentals. Many of the better books were long out of print: Langhaar's *Dimensional analysis and theory of models*. (Chapman & Hall (1951)) is one of them. Birkhoff's *Hydrodynamics: a study in logic, fact and similitude* (Princeton University Press (1960)) is an inspiration, and everyone ought to read it at some time, but as an introduction for undergraduates it is inaccessible and overdemanding.

Students coming from school often undervalue dimensional analysis and are confused about what it can accomplish. They imagine that dimensional analysis can be used to confirm results that have been secured by some other route, but that it cannot be used to derive new results. Often the method has been linked to power-law relationships, and they suppose that all relationships have that form, or that all relationships to which dimensional analysis applies are necessarily power laws, or that dimensional analysis is part of fluid mechanics and has no relevance anywhere else. Those notions are all completely false, and can be dangerously misleading.

I had in mind a slim volume that would put forward the basis ideas succinctly but accurately, and that would include many examples from engineering and physics, carefully chosen so that they are interesting and sometimes surprising. The starting point was a conversation with my friend and colleague Rex Britter. At first Rex was going to be a co-author, but in the event his many other commitments made that impracticable for him, something that I much regret. He showed me notes on dimensional analysis written for students by the late Harry Shercliff, who for a lamentably brief time was Head of the Department of Engineering at Cambridge, and before that a professor at Warwick: they influenced the approach to the underlying theory in Chapter 4. Herbert Huppert of the Department of Applied Mathematics and Theoretical Physics gave a lecture for schools, with the engaging subtitle 'something for (almost) nothing', and that was another inspiration.

I am grateful to the following people for various kinds of encouragement, help, and inspiration: Jack Apgar, Holger Babinsky, the late John Baker, Bob Brown, Chris Calladine, Bill Dawes, the late Dan Drucker, the late Jack Ells (who gave me two of the problems discussed in Chapter 4), John Halkyard, Jacques Heyman, Stephen Huntington, Hamid Jafar, Roger King, Ibrahim Konuk, Leng Shuh Pei, the late Douglas Maclellan (who taught me about dimensional analysis as an undergraduate), James Martin, the late Tom McMahon, Allan McRobie, Caroline Michel, Brian Pippard, Alan Reece, Tim Sanderson, Andrew Schofield, Susan Sterrett (who has written a fascinating book on the links between dimensional analysis and philosophy), Milton Van Dyke, David Walker, Wang Chien Ming, and Derek Yetman.

The book was begun at my home in Maine, and completed at Cambridge University and at the National University of Singapore. I am grateful to those institutions for intellectual companionship, for the opportunity to write, and for their excellent library facilities, and in particular to Seeram Ramakrishna, Chan Eng Soon, and Choo Yoo Sang for inviting me to Singapore. I would like to thank Gow Huey Ling, Tan Yi Xin, and their colleagues at World Scientific for the many ways in which they helped toward the completion of the book.

My beloved wife Jane has helped me immeasurably through her encouragement, reassurance, and almost infinite patience.

The mistakes are mine, and I would like to know about them.

Andrew Palmer
Singapore, March 2007

Contents

Chapter 1

Introduction

1.1 Motivation: how to get something for nothing

Why is dimensional analysis worth studying ?

Dimensional analysis is a technique that can be used to clarify and explain relationships between physical quantities. The idea that it exploits is that fundamental relationships between physical quantities must relate quantities of the same kind, and cannot mix up quantities of different kinds. We can say that the distance from Cambridge to London is 50 miles, or that it is 80 km, or (if we wish) that it is 8×10^7 mm or 5.28×10^{-12} light years, but we cannot meaningfully say that the distance is 20 K or 10°C. Units like miles, kilometers, and kilograms tell us what kind of quantity we are dealing with. 'Dimension' is a slightly more formal way of denoting that familiar idea.

The units are themselves a human construct, but the fundamental relationships in the physical world can have nothing whatsoever to do with human beings. They have existed from the beginning of time, and they will exist long after humanity has departed. If we measure the time it takes for a pendulum to swing back and forth, and find it to be 3.7 seconds, the second is a unit constructed by people, with a history long in human terms (with influences going back at least to the Babylonians, who counted in sixties), and the 3.7 is our measurement of the time in those units that we defined. *The pendulum itself has nothing to do with the Babylonians, with us, or with anyone in between.* The time it takes to swing is determined by the fundamental mechanics, and depends on quantities like the length, the acceleration due to gravity in the particular place it is sited, the mass, and how far it swings. A pendulum described by the same quantities and set up a billion years in the past, or a billion years in the future, would take just the same time to swing: it would know nothing of seconds. Likewise, it would know nothing of particular numbers like 3.7, which depend on how many fingers and toes we happen to have.

At first it seems unlikely that such a simple idea would be able to tell us much, but it turns out to be unexpectedly fruitful. Rayleigh put it well:

> *"It happens not infrequently that results in the form of 'laws' are put forward as novelties on the basis of elaborate experiments, which might have been predicted a priori after a few minutes consideration"*[1]

Had he been writing today, he might have added after 'experiments' the words 'or numerical computations.' A less elegant way of phrasing that is that dimensional analysis can give us something for (almost) nothing. Some quiet reflection will often save us a great deal of work.

Dimensional analysis does not usually give us a complete analysis, but it provides useful steps toward a complete understanding. To continue the simple example of a pendulum, it is not immediately obvious that the natural period of a pendulum's swing is proportional to the square root of the length of the pendulum. Dimensional analysis will show us that it must be. It does not tell us the constant of proportionality, but that we can find by experiment or by a more complete analysis.

Dimensional analysis can also be used to assemble the results of experiments in a concise and accessible form, so that we can arrive at broadly applicable general results from a small number of tests, often at a model scale. For example, engineers frequently want to transmit fluids along pipes, and need to know the pressure difference required to generate a required flow rate. The pipes have very different diameters, from several meters (penstocks in hydroelectric plants, magma channels in volcanoes) down to small fractions of a millimeter (pores between sand particles in an oil reservoir in rock, blood vessels in the brain, microchannels in nanofluidic devices), and the fluids are very different (air, water, molten rock, gas, oil, liquid sodium, sewage). It would be laborious and awkward if every design calculation for a fluid in a pipe had to be based on a previous experiment on a pipe with the same fluid and the same diameter. Dimensional analysis tells us that we can carry out experiments with one diameter and one fluid at different velocities, and that if we analyze the results in the right way we can derive a universal curve which we can apply to any diameter and almost any fluid.

Bad teaching sometimes leads students to misconceptions about dimensions and analysis based on them. They imagine that the methods can only be used to confirm results that are already known, and that nothing new can emerge. That notion is completely bogus. Some students also think that dimensional analysis is used only in fluid mechanics. The objective of this book is to show that dimensional analysis is a powerful and broad-ranging tool that will give us new results that are not obvious. I do not ask that to be taken on trust, and shall demonstrate it by example.

The first example follows: it is deliberately elementary and separate from physics and engineering.

1.2   Pythagoras' theorem

Asked to cite a mathematical theorem that has someone's name attached to it, most people choose Pythagoras' theorem.

Many of those people could state the theorem correctly. It is about right-angled triangles, and it says that the square of the length of the longest side of the triangle is equal to the sum of the squares of the lengths of the other two sides. *Square* means the length multiplied by itself, so if one of the lengths is 3 length units the corresponding square is $3 \times 3 = 9$ units squared. 3 squared is written 3^2. The simplest and best-known triangle that satisfies the theorem has the lengths of its sides in the ratio 3:4:5, and

$$3^2 + 4^2 = 9 + 16 = 25 = 5^2$$

The next simplest has sides in the ratio 5:12:13, and

$$5^2 + 12^2 = 25 + 144 = 169 = 13^2$$

These are special cases: it happens that each of the lengths is a whole number, measured in whatever units we chose, but the result does not depend on any of the lengths being a whole number.

Few people nowadays could prove Pythagoras' theorem by the traditional method of Euclid's geometry. In the future there will be fewer still, as the notion of formal proof falls out of fashion in mathematical education. Instead of applying Euclid, we can prove the theorem by dimensional analysis.

Consider two squares, one with sides twice as long as the other (Fig. 1.1a). The area of the large square is four times the area of the small one: if we doubt that, we can place four of the smaller squares within the large one, without overlapping them and without leaving any space uncovered (Fig. 1.1b). Four is the square of two, two times two. Similarly, if we have two squares, one with sides three times as long as the other (Fig. 1.2a), the area of the large square is nine times the area of the small one (Fig. 1.2b). Nine is the square of three, three times three. In general, the area of a square is proportional to the length of one side multiplied by itself, that is, to the square of the length of one side.

Squares all have the same shape, with four equal sides and four equal angles. Triangles do not all have the same shape, and in general we need two angles to define the shape of a triangle: geometry calls triangles with the same shape *similar* triangles. A right-angled triangles has one 90° angle and two smaller angles. Not all right-angled triangles have the same shape. However, we can define the shape unambiguously if we define one of the two smaller angles. All right-angled triangles with one angle 35° (say) have the same shape, and Fig. 1.3a shows three of them.

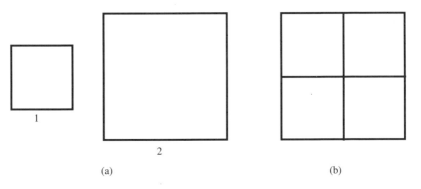

1

2

(a)   (b)

Fig. 1.1. Squares.

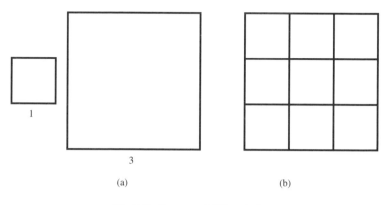

Fig. 1.2. Squares of different sizes.

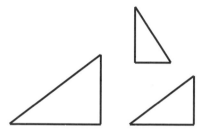

Fig. 1.3. Right-angled triangles with smallest angle 35°.

One of the smaller angles completely defines the shape of a right-angled triangle. The length of one of the sides defines how large the triangle is. So as to eliminate confusion as to which angle and which side are being referred to, think of the angle that defines the shape as the smallest angle, and think of the longest of the three sides as the one that defines the size: geometry calls the longest side the *hypotenuse.*

If two right-angled triangles have the same shape, and one has a longest side twice as large as the other, the second triangle has four times the area of the first (Fig. 1.4a), and we can check that by placing four of the small triangles on top of the large one. They fit without overlapping or leaving gaps (Fig. 1.4b). If two right-angled triangles have the same shape, and one has a longest side three times as large as the other, the second triangle has nine times the area of the first, and so on. It follows that the area of right-angled triangles with the same shape is proportional to the

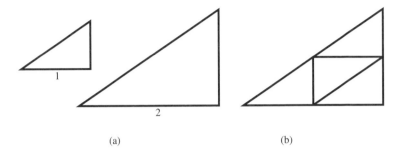

(a)   (b)

Fig. 1.4. Triangles of different sizes.

square of the length of the longest side. 'With the same shape' is important: right-angled triangles with the same longest sides but different shapes have different areas (Fig. 1.5).

The area of a right-angled triangle is therefore proportional to the square of the longest side multiplied by something that depends on the smallest angle. Call the longest side s and the smallest angle a. The area of the triangle is

$$\text{area} = (\text{longest side } s)^2 f(\text{smallest angle } a) \qquad (1.2.1)$$

f(smallest angle) means a quantity that depends on the smallest angle, in mathematical language a function of the smallest angle. The function f reflects the fact that the shape matters, but for the present purpose all we need to know about the form of the function is that it is single-valued and not zero. 'Single-valued' means that for one value of the smallest angle there is one and only one value of f(smallest angle). The function is not

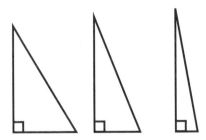

Fig. 1.5. Triangles of different shapes.

zero unless the smallest angle is zero, in which case there is no triangle and therefore no area.

Now think of a general right-angled triangle ABC (Fig. 1.6a), with the smallest angle at A and the right angle at C, and draw a line from C at right angles to the longest side AB (Fig. 1.6b). The line reaches the longest side at D. Call the smallest angle a, the angle at corner A. a is the smallest angle in the right-angled triangle ABC, whose longest side is AB. a is also the smallest angle in the right-angled triangle ACD. The angle ACD is $90° - a$, because the three internal angles in the triangle ACD must add up to $180°$. The angle BCD is the right angle BCA minus ACD, and so is $90° - (90° - a)$ and therefore a: that is added to Fig. 1.6c.

Now consider the areas of the triangles, using the general relationship (1.2.1) to work them out. Triangle ABC has its longest side AB and its smallest angle a, and its area is

$$\text{area ABC} = AB^2 f(a) \tag{1.2.2}$$

Triangle ACD has its longest side AC and its smallest angle a, and its area is

$$\text{area ACD} = AC^2 f(a) \tag{1.2.3}$$

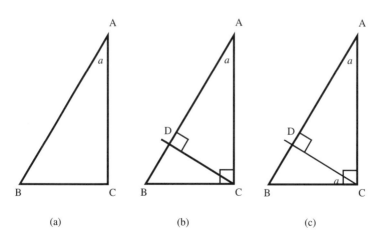

Fig. 1.6. Right-angled triangles.

Triangle BCD has its longest side BC and its smallest angle a, and its area is

$$\text{area BCD} = \text{BC}^2 f(a) \tag{1.2.4}$$

The area of ABC is the sum of the area of ACD and the area of BCD:

$$\text{area ABC} = \text{area ACD} + \text{area BCD} \tag{1.2.5}$$

and so, substituting the equations for each area

$$\text{AB}^2 f(a) = \text{AC}^2 f(a) + \text{BC}^2 f(a) \tag{1.2.6}$$

Now divide by $f(a)$, which we can safely do because it is not zero, and then we get

$$\text{AB}^2 = \text{AC}^2 + \text{BC}^2 \tag{1.2.7}$$

which is Pythagoras' theorem for right-angled triangles. In words, the square of the length of the longest side is equal to the sum of the squares of the lengths of the two shorter sides.

What we call Pythagoras' theorem is not actually due to Pythagoras, and was known to the Babylonians a thousand years before Pythagoras lived.[2] It is dangerous to speculate about how mathematical results were derived in antiquity, because people then were no less intelligent than us but had different contexts, education, priorities, and frames of reference. Still, we do know that the Babylonians were much concerned with measuring land, and had clear ideas about area and volume. We also know that the Babylonians often used geometrical arguments to solve problems that we usually think of as algebraic: for example, they solved the quadratic equation $x^2 + x = A$ by looking for a rectangle with area A and one side unit length longer than the other.[3] We can pause to think of the theorem being first derived by a Babylonian land-measurer sitting under a tree on a hot afternoon. Our word 'geometry' comes from the Greek words 'ge' ($\gamma \acute{\eta}$) 'earth' and 'metria' ($\mu \varepsilon \tau \rho \acute{\iota} \alpha$) 'measuring'.

The derivation intentionally does not use the formalism of dimensional analysis, but all the same it is essentially a dimensional argument, because the key idea is that the areas of triangles that have the same shape are proportional to the squares of the lengths of the longest sides. It demonstrates that dimensional analysis can be applied to derive useful results that are not trivial and not obvious.

A way forward into dimensional analysis proper is to start by thinking about units, and we do that next.

References

1. Rayleigh L (1915). The principle of similitude. *Nature*, 95, 66–68.

2. Neugebauer, O (1957). *The Exact Sciences in Antiquity.* Providence, RI, Brown University Press.

3. Høyrup, J (1994). Babylonian mathematics. In *Comparitive Encyclopaedia of the History and Philosophy of the Mathematical Sciences*, Vol. 1, pp. 21–30, London: Routledge.

Chapter 2

Numbers and Units in Engineering and Physics

2.1 Introduction

Scientists use numbers in different ways. Sometimes they are simply iden-tifiers that happen to be in the form of numbers. Sometimes they are scales, which are linked to a rank order but not to a unit. More often they refer to quantities measured in some units.

Engineering and science attempt to construct useful relationships between different kinds of quantities. At the simplest level, there is a rela-tionship between the dimensions of a body, a material property that we call density, and the mass of the body. Similarly, there is a relationship between the mass of the body, the acceleration due to gravity, and the weight of the body. At a more complex level, the force needed to propel a ship through a smooth sea depends on the dimensions of the ship, the density of the water (and the air), and the viscosity of the water (a measure of how easily it deforms: honey has a higher viscosity than water). If the sea becomes rough, the height, length, and periods of the waves become important.

Some of those relationships apply only to specific choices of units, but some are "unit-free" and apply in any units. Dimensional analysis is principally concerned with the last class of relationships, and they often turn out to be the most powerful. But first we consider other kinds of relationship.

2.2   Labels

Numbers often simply label and identify things. The telephone dialing code for Cambridge is 1223, and the code for Aberdeen is 1224, but 1223.5 may not be a code at all, and it certainly does not signify somewhere halfway between Cambridge and Aberdeen. Numbers that are really just labels could be replaced by letters, or by hieroglyphs or Chinese characters, or in some contexts by musical tones, smells, or colors.

In principle, we could identify physical quantities in the same way. Having arrived at the concept of length, and having a way of classifying lengths as the same or different, we could label one group of identical lengths with the number 1224, and label another group of identical but quite different lengths with the number 1223, and so on. The same idea could be applied to mass and time. It might just be possible to develop geometry and mechanics on that basis, but it would be extraordinarily cumbersome and unrevealing.

2.3   Scales

The next concept is a scale unrelated to a unit.

Mariners describe the roughness of the sea by a 'sea state' scale from 0 (flat calm) to 12 (hurricane). The verbal descriptions are laid down by the World Meteorological Organization. Sea state 4 is described as "small waves, becoming longer; fairly frequent white horses" (and in the racing sailor's description by Marchay as "great pleasure"). Sea state 8 is described as "moderately high waves, edges of crests begin to break into the spin-drift" (Marchay; "fear tinged with terror"). Larger numbers in the scale correspond to rougher seas, but it is not intended that sea state 8 is in any sense 'twice' sea state 4. The scale can be learned by experience at sea, and by visual comparison between the sea and standard photographs.[1] Sea state is related to wind, but is not the same: there can be rough seas with almost no wind, or for brief periods high winds with hardly any waves.

Scales that have no units can be very useful for many purposes, and may form a workable description of complex situations. An offshore construction specialist knows well what it means to say that an operation requires a weather window in which the sea state is lower than 4 for at least 12 h. It is possible to go further, and to relate sea state to a more quantitative measure such as significant wave height (the average crest-to-trough height of the highest one-third of the waves), but for many purposes that is unnecessary.

Another example of a useful scale without a unit is Moh's scale of hardness, which runs in sequence from softer to harder minerals, from 1 for talc up to 10 for diamond. On that scale fingernail has a hardness of about 2.5, halfway between 2 for rock salt and 3 for calcite. A fingernail is therefore harder than rock salt, and can scratch its surface, but not as hard as calcite, which it will not scratch. The scale is useful in practice, for instance in a geologist's field identification of rocks, and the absence of a unit does not detract from its value.

Two more examples are wire gauge and the notoriously controversial 'intelligence quotient' (IQ). There are at least four scales for wire: they are based on diameter, but not in linear way, so that in the gauge series most commonly applied in the United Kingdom the difference between gauges 20 and 21 is 0.101 mm in diameter but the difference between gauges 30 and 31 is 0.020 mm. IQ is a ratio between the score on a standardized test and an average of 100.

2.4 Measurement with units

A much more broadly applicable approach to physical quantities is the idea of a measure. Imagine that points A, B, and C lie in a straight line. We lay a measuring stick on the line with one end at A, mark the point at the other end of the stick K, lay the stick down again with one end at K, mark the other end L, and so on. If we repeat the step three times (say), we come to B. If we start at B and continue along BC, we have to lay the stick down five times to get from B to C. We then know that if we start at A, we will have to lay the stick down $3 + 5 = 8$ times to come to C. The length AB is 3 units, and the length BC is 5 units, and the length AB and BC together is 8 units.

A length we wish to measure in this way may not correspond to a whole number of stick lengths: 7 might be too few, and 8 too many. Then we can subdivide the measuring stick into tenths, say, and assign a meaning to 7.4 stick lengths, or go further and subdivide each tenth into tenths, and so on.

A similar process can be used to measure volume. We want to have a measure of the volume of a bucket, and we have a cup. We fill the cup to the brim with water, pour it into the bucket, fill it a second time, pour that water into the bucket, and so on until the bucket is full, counting how many cupfuls we need. The count is the volume of the bucket in cups. If we have a second different bucket, the volume of the two buckets together is the sum of the volumes of the buckets separately.

Most useful physical units have this property. Quantities measured in these units can be added and multiplied: three units of something have the properties of three times as much of that something as one unit does. We are so accustomed to measuring in units that allow us to take these steps that we forget that other ways could be used. This kind of measurement is the basis of Kelvin's famous statement that unless we can measure something our knowledge is of an insubstantial kind.

Units for length, mass, time, velocity, electrical resistance, and so on are all additive. Not quite all units are additive: there are units like pH and decibels that are logarithms of ratios (often referred to standard values measured in terms of more conventional units).

2.5   Standardized units

Someone isolated on a desert island could carry out the processes described in Sec. 2.4. He would use his own measuring stick and measuring cup, and his measurements would be in those units. But then if he wanted to describe his measurements to somebody on another island, he would have to have some way of relating his stick and cup to the other person's different stick and cup. Communication becomes possible if measures are somehow standardized, so that measurements on one island can be related to measurements on another island. The units do not have to be the same, but there has to be a relation between them, in some such terms as 'my measuring cup is one-and-a-third times as large as yours.'

Efforts to standardize measures have a fascinating history over thousands of years. Many units were originally related to parts of the body. Cloth was measured in ells: one ell is the length of a forearm from hand to elbow. The Dutch and French words for 'inch' are 'duim' and 'pouce,' the words for thumb. A horse's height is still measured in 'hands.' But there is of course variability between people, and so more precise standards were made. The English yard is supposed originally to have been the distance between Henry I's nose and the thumb of his outstretched arm, but it is plainly impracticable to have to keep going back to that standard, and so it was translated into a reference standard length of metal.

Units are nowadays defined either by their relation to natural properties or to arbitrary material standards. The meter was originally defined as one forty-millionth of the circumference of the Earth, in keeping with the spirit of the French Revolution and its rejection of the royalist past. Then from 1889 to 1960 it was the distance between two lines engraved on a platinum–iridium bar at the International Bureau of Weights and Measures,[2] at Sèvres outside Paris. Now it is 1,650,763.73 wavelengths in vacuum of the radiation corresponding to the transition between levels $2p_{10}$ and $5d_5$ of the kryton-86 atom. Mass, on the other hand, is still related to an arbitrary physical reference mass, a platinum–iridium cylinder kept at Sèvres.

Imperial units were once based on physical standards, but are now related back to metric standards: the inch is 0.0254 m exactly.

2.6   Converting from one unit to another

All science and most engineering (except in the United States) are based on metric units, nowadays almost always through the SI (Systeme International) system agreed at the 11th General Conferences on Weights and Measures in 1960.[3]

The unit of length is meter (m), the unit of mass is kilogram (kg), and the unit of time is second (s). There is a standardized notation: units that commemorate individuals have capitals (e.g., Pa for Pascal, the unit of pressure, after Blaise Pascal (1623–1662)), and there are standard prefixes for multiple and submultiple units, such as M ('mega') for million and G ('giga') for billion, and rules about not putting in stops for abbreviations. The unit of force is Newton (N).

An important feature is that SI units are consistent: one unit of force acting on one unit of mass causes the mass to accelerate with one unit of acceleration. That is not true in some other systems.

We still need to be able to convert from one unit to another, for three reasons. First, some older multiple units persist, and they may be convenient and useful. We could in principle measure long times in kiloseconds and gigaseconds, but the minute (60 s), hour (3600 s), and day (86,400 s) are almost always more convenient. Mariners measure speed in knots (nautical miles/hour) and the oil industry measures oil in barrels: neither is likely to give them up. Second, until about 30 years ago much engineering in English-speaking countries was based on the Imperial (foot pound second) system. Some industries hang on to parts of that system, engineering in the United States still uses much of it with local variations, and we need to be able to make use of data in Imperial or American units. Third, some nominally metric countries have not abandoned units that are metric but not SI: German and French engineers still use the kilopond (kilogram force) and the metric horsepower.

The most straightforward and mistake-proof way of converting from one unit system to another is to use the fact that if we multiply a quantity by a ratio identically equal to 1 we do not change the quantity. One hour is 3600 s. The ratio (3600 s/1 h) is therefore identically 1. Accordingly

$$1\,h = 1\,h \times \frac{3600\,s}{1\,h} = 3600\,s$$

The same idea can be used to convert composite units such as speed. Suppose that we want to convert speed in ft/s to m/h. One foot is 0.3048 m, and so (0.3048 m/1 ft) = 1. One hour is 3600 s, and so (3600 s/1 h) is 1. Therefore

$$1\,ft/s = \frac{1\,ft}{1\,s} \times \frac{0.3048\,m}{1\,ft} \times \frac{3600\,s}{1\,h} = 1.09728 \times 10^3\,m/h$$

where hour and second have been treated as algebraic quantities and cancelled between the nominator and the denominator.

Similarly, if we wish to convert volumetric flow rate from m^3/h to ft^3/s, we use the identities

$$1\,ft = 0.3048\,m$$
$$(1\,ft)^3 = (0.3048\,m)^3$$
$$1\,h = 3600\,s$$

and so

$$\frac{1\,m^3}{1\,h} = \frac{1\,m^3}{1\,h} \times \frac{1\,ft^3}{(0.3048\,m)^3} \times \frac{1\,h}{3600\,s} = 9.810 \times 10^{-3}\,ft^3/s$$

More complex compound units transform in the same way. Thermal conductivity is the measure of how easily a material conducts heat. The rate of flow of energy, measured as energy per unit area per unit time, is proportional to the temperature gradient, measured as rate of change of temperature with distance. Energy flow rate is measured in watts (W). One W is 1 J (joule) per second, and 1 J is 1 N m. Heat flow rate per unit area is therefore measured in W/m^2 (watts per square meter). Temperature gradient is measured in $°C/m$ (degrees Celsius per meter). Thermal conductivity is the ratio between energy flow rate per unit area and temperature gradient, and its unit is

$$\frac{W/m^2}{°C/m} = W/m\,°C$$

British practice in the past used more than one unit for thermal conductivity, but the most common was Btu/ft h °F. Btu stands for British Thermal Unit (the energy required to raise the temperature of 1 pound of water by 1 °F): 1 Btu is 1055.06 J. One foot (ft) is 0.3048 m, one hour (h) is 3600 s, and one degree Fahrenheit is 1/1.8 °C. Following the same procedure as before, and always multiplying and dividing by factors that we know to be unity, such as $\left(\frac{1°C}{1.8°F}\right)$, the conversion factor is

$$1\frac{Btu}{ft\,h\,°F} = 1\frac{Btu}{ft\,h\,°F} \times \frac{\left(\frac{1055.06\,J}{1\,Btu}\right) \times \left(\frac{1\,W}{1\,J/s}\right)}{\left(\frac{0.3048\,m}{1\,ft}\right) \times \left(\frac{3600\,s}{1\,h}\right) \times \left(\frac{1°C}{1.8°F}\right)}$$
$$= 1.73073\ W/m\,°C$$

Some nonmetric force units involve the standard acceleration of gravity, which is why they often lead to mistakes and confusion, and ought to be avoided wherever possible. One lbf (pound force) is the weight of 1 lb mass, which is 0.45359237 kg (exactly, because nowadays the pound is legally defined in terms of the kilogram). The standard acceleration of gravity is $9.80665 \, \text{m/s}^2$, and so the weight of 1 kg is $1 \, \text{kg} \times 9.80665 \, \text{m}^2/\text{s} = 9.80665 \, \text{kg m/s}^2 = 9.80665 \, \text{N}$. Accordingly

$$1 \, \text{lb f} = 0.45359237 \, \text{kg} \times 9.80665 \, \text{m}^2/\text{s} = 4.448 \, \text{N}$$

Dimensional analysis can be thought of as a generalization of these ideas, but in Chap. 3 we first revisit the idea of variables.

2.7   Problems

2.1   Complete the following:

1 year	= ....s (time)
1 chain (22 yards, 66 ft)	= ....m (length)
1 mile (1760 yards, 5280 ft)	= ....m (length)
1 acre (4840 square yards)	= ......m^2 (area)
1 acre ft	= ....m^3 (volume)
$1 \, \text{ft}^3/\text{min}$	= ......m^3/h (volumetric flow rate)
1 mile/h	= .... m/s (velocity)
1 knot (1 nautical mile/h)	= .... m/s (velocity)
1 UK ton (2240 lb)	= .... kg (mass)
$1 \, \text{kg/m}^3$	= ......lb/ft^3 (density)
$1 \, \text{lbf/in}^2$	= ....Pa (pressure)
$1 \, \text{lbf/in}^2$	= ......MPa (pressure)
1 mm Hg	= .... Pa (pressure)
1 horsepower (550 ft lbf/s)	= ....W (power)
1 tonne force (2000 kg force)	= ....N (force)
1 lb/ft	= ....kg/m (mass per unit length)

The chain is an old British unit, and is the length of a cricket pitch. The acre foot is a unit of water volume used in irrigation for agriculture, and is the volume of water that covers 1 acre to a depth of 1 ft. One nautical mile is

1853.18 m in the United Kingdom but 1852 m exactly everywhere else. The density of mercury is 13,590 kg/m^3.

2.2 Body mass index (BMI) is a measure of how thin or fat someone is. It is defined as the person's weight divided by the square of his or her height. If weight is measured in kilograms (kg, strictly speaking kgf) and height in meters, a person is supposed to be obese if his or her BMI is greater than 30. What is the corresponding number if weight is measured in pounds and height in feet?

BMI ought strictly to be defined in terms of mass rather than weight: if an astronaut goes to the moon, his weight is much smaller but his BMI is the same. However, the definition in terms of weight is sensible in practice, because it is important to make the definition accessible to as many people as possible, and not everyone understands — or needs to understand — the distinction between weight and mass.

2.3 An old British unit for thermal conductivity is

Btu/h/sqft/cm/°F

quoted in Ref. 4, with derisive intent, and supposedly useful for calculating the heat transmission of walls. Derive the conversion factor between that unit and the SI unit. A Btu (British Thermal Unit) is the quantity of heat needed to raise the temperature of one pound mass of water by 1°F. The specific heat of water is 4185.5 J/kg °C.

2.4 Fuel consumption of cars in the United Kingdom used to be measured in miles per gallon, but is now measured in km/liter. One UK gallon is 0.00454609 m^3. One liter is 0.001 m^3: the liter is not an SI unit, and very slightly different definitions have been applied, so its use is discouraged for precise measurement, but for this purpose it is adequate. Derive the conversion factor between miles per gallon and km/liter.

2.5 At the time of writing, the pump price of petrol (gasoline) in the United States is 3.24 US$/US gallon. One US gallon is 0.8 UK gallons, and £1 is worth US$2.08. Determine the United States price in £/liter (and in your own currency if that is different) and compare it with the price where you live.

2.6   Oil is transported across country by pipelines. The choice of the diameter of an oil pipeline has to be based on a trade-off. Increasing the diameter increases material and construction costs, but reduces the pressure drop along the line and therefore reduces the power consumed in pumping. Reducing the diameter increases the pressure drop and the pumping power, but reduces construction costs. A widely applied approximate formula for the optimal diameter is, in American units:

$$D = \sqrt{(q/500)}$$

where D is the diameter in inches and q is the design maximum flow rate in barrels/day (1 barrel = 42 US gallons = 0.158987 m^3).

Most of the world uses metric units. Find an equivalent formula with D measured in mm and q in m^3/s.

References

1. Meteorological Office State of Sea Booklet (1993). London: Her Majesty's Stationery Office.

2. Bernardis, M-A and B Hagene (1995). *Mésures et Démesures*. Paris: Editions de la Cité des Sciences et de l'Industrie.

3. Changing to the metric system (1967). London: Her Majesty's Stationery Office.

4. Weber, RL (ed) (1973). A random walk in science. London: Institute of Physics.

Chapter 3

Dimensions, Dimensionless Groups, and Variables

3.1 The concept of dimension

The idea of dimension is a development of the ideas that have already been used to convert units. **Dimensions are just a code that tells us how the numerical value of a quantity changes when the units of measurement change.**

Some authors ascribe a deeper pseudo-philosophical significance to dimensions. None of the applications of dimensional analysis rely on that, and here we adopt a pragmatic approach. At one time there was a lot of discussion about the true dimensions of different quantities, but that discussion leads nowhere. The physicist Planck is supposed to have remarked that to ask for the "true" dimension of a quantity was as pointless as to ask for the "true" name of something.

In Chap. 2 we converted $1\,h$ to seconds by applying the identity $1\,h = 3600\,s$ and writing;

$$1\,h = 1\,h \times \frac{3600\,s}{1\,h} = 3600\,s$$

We did that because we recognized $1\,h$ as a time. We say that its dimensions are time, denoted T, and often we indicate that we are talking about dimensions by writing T in square brackets, thus: time has dimension [T].

It would have made no sense to attempt to convert 1 h to seconds by applying the different identity 1 foot = 304.8 mm and writing

$$1\,h = 1\,h \times \frac{304.8\,mm}{1\,foot} = ?$$

The identity 1 ft = 304.8 mm is perfectly correct, but it refers to a qualitatively different quantity, length. Lengths are different in kind from times, and we cannot convert a time into a length or a length into a time. But if we know that we are dealing with lengths, then we can make use of 1 ft = 304.8 mm, and of other unit conversions that apply to lengths, and write

$$1\,mile = 1\,mile \times \frac{5280\,ft}{1\,mile} \times \frac{304.8\,mm}{1\,ft} = 1{,}609{,}344\,mm$$

We say that lengths have dimension [L].

We could do the same with composite units. Velocity is distance traveled in unit time. One mile/h is 447.04 mm/s, and so we could say

$$1\,mile/h = 1\,mile/h \times \frac{447.04\,mm/s}{1\,mile/h} = 447.04\,mm/s$$

and assign velocity another independent dimension [V]. Generally, though, it is most instructive to reduce the number of dimensions as far as possible, and to work in terms of a small number of basic dimensions. Velocity is a ratio of length to time, and so we can use the fact that length and time have their own independent dimensions. We can use the unit conversions that belong to those dimensions, and write instead

$$1\,mile/h = \frac{1\,mile}{1\,h} \times \frac{\frac{1{,}609{,}344\,mm}{1\,mile}}{\frac{3600\,s}{1\,h}} = 447.04\,mm/s$$

The conversion factor for length [L] appears in the numerator (top) of the large fraction, and the conversion factor for time [T] appears in the denominator (bottom) of the large fraction. In terms of the dimensions [L] and [T], the dimension of velocity is [[L]/[T]] , which is more easily written as [LT^{-1}].

Dimensional analysis usually takes the basic dimensions to be:

mass	M
length	L
time	T
temperature	Θ
electric charge	Q

and determines the dimensions of other quantities in terms of those fundamental dimensions. The dimension of velocity is that of length divided by time, denoted $[LT^{-1}]$. Acceleration is rate of change of velocity with time, and so the dimension of acceleration is that of velocity divided by time, denoted $[LT^{-1}]/[T]$ or $[LT^{-2}]$. Force is mass times acceleration $[M] \times [LT^{-2}]$ $= [MLT^{-2}]$. Energy or work is force times distance $[ML^2T^{-2}]$, and so on. Temperature, though, is another different kind of quantity: we cannot manipulate combinations of mass, length, and time to arrive at temperature. The same applies to electric charge.

There is nothing magic or special about the choice of M, k, L, T, θ, and Q as fundamental units, although they are often the most convenient. We can make alternative choices and get the same results, as we shall see later. The example above suggested that in principle we can give a composite unit a dimension of its own. Often that is counter-productive, but sometimes we can do so usefully. Dimensional analysis ought not to be thought of as a set of rigid rules, or as a machine-like process. We can use it most fruitfully as part of a process of exploration and search, in which there are no firm rules.

Pure numbers are dimensionless, and are denoted [0]. Angles too have dimension zero: we can think of them as a ratio between two lengths, a radius and an arc length, and their dimension is therefore $[L]/[L] = [0]$.

3.2 Dimensional consistency

Units are a human construct, as Chap. 2 described. Units referred to standards such as the mass of a lump of platinum in France or the distance between Henry I's nose and his fingers are plainly arbitrary and artificial.

Fundamental physical relationships have a higher standing. They describe something that underlies the way the world is put together. They have nothing to do with human beings, and they existed before human beings and will still exist after them.

The most fundamental relationships between quantities are dimensionally consistent and unit-free. If we express them as equations, dimensional consistency means that both sides of the equation have the same dimensions. For example, if

$$x = y + z$$

then x, y, and z must all have the dimension [L], or all have the dimension $[MLT^{-1}]$, or all be dimensionless, and so on. A dimensionally consistent equation cannot have x with dimension [L] but y and z with dimension $[MLT^{-1}]$. Only dimensionally consistent equations can be *unit-free* and apply in any units: we can demonstrate this by taking an inconsistent relationship and trying to convert its units. A relationship can be dimensionally inconsistent but very useful all the same, and some examples are given in Sec. 3.3.

We could combine dimensionally consistent relationships to make relationships which are not consistent. If

the cost of a bag of apples $= y \, £$

and

the distance from Glasgow to Edinburgh $= z \, km$

then we could add the two equations and say

the cost of a bag of apples + the distance from Glasgow to Edinburgh
$$= y + z$$

or we could multiply the first equation by any number a and the second by any other number b, and add them to get

a(the cost of a bag of apples) $+ b$(the distance from Glasgow to Edinburgh
$$= ay + bz$$

for any a and any b. Relationships of this kind are formally correct but not useful, because different kinds of quantities are mixed up. We can only make them useful by splitting them back into dimensionally consistent parts, which we can do by setting a to 0 or b to 0 and getting back the original equations.

3.3 Empirical relationships that are not dimensionally consistent

Some empirical relationships are useful but not dimensionally consistent, and only apply to some particular units or under some particular conditions. For example:

1. Crickets

In many parts of the United States, sunny days in summer are loud with the cheeping noises of crickets. The air temperature θ in degrees Fahrenheit is supposed to be related to the cheep frequency by

$$\theta = 37 + (\text{number of cheeps in 15 seconds})$$

2. Hurricanes

At the eye of a hurricane, the wind speed is quite low. A long way from the hurricane the speed is low. Somewhere in between the wind speed is a maximum. The distance from the center of the eye to the location where the wind speed is highest is given by:

$$R_m = 1.007(\text{LAT}) - 0.187(\Delta p) + 9.806$$

where R_m is the radius to the maximum wind speed, in nautical miles, LAT the latitude in degrees, and Δp is the difference between the lowest pressure at the hurricane center and the pressure far from the center, in hPa (millibars).

3. Optimal diameter of oil pipelines

A pipeline is designed to carry a specified fluid flow rate between two points. If the pipeline is small, there will be a large pressure drop between one end and the other, a large amount of power will be consumed in pumping, and the pumps will be expensive. If the pipeline is large, the pressure drop is smaller, the pumps can be smaller and will need less power, but the pipeline itself will be more expensive to construct. Somewhere in between there is an optimal diameter that minimizes total costs. A useful rule of thumb[1] for oil pipelines says that the optimal diameter is approximately given by

$$D = \sqrt{\frac{q}{500}}$$

where D is the diameter in inches and q is the flow rate in barrels/day.

4. Marriage

Pippard[2] quotes an "ancient rule" that the optimal ages of a man (m) and a woman (f) who marry are related by

$$f = (1/2)m + 7$$

Formulas of those kinds can be very useful. They are not to be despised in the right context, and form a useful part of the practice of technology. They are approximate statements that summarize human experience of extremely complex situations. The first example is interesting to someone walking in the forest, the second to a tanker captain sailing toward the path of a storm or to a fire chief planning disaster response, the third to an engineer estimating the cost of developing an oil field, and the fourth — if has any value at all! — to an unusually cold-blooded couple nervous of being guided by passion alone.

    The hurricane example involves many issues to do with the dynamics of hurricane evolution, including the Coriolis acceleration of a revolving fluid system on a rotating earth (which is why the latitude is relevant). The pipeline example 3 involves factors such as the price of steel, the price of pumps, the viscosity of the oil, the value of the power used to drive the pumps, and the cost of clearing the pipeline route and welding the pipe. An exact answer would be different in Texas and Arabia, and would be different in 2007 and 2017. None of the contributing factors are explicitly included. Similarly, example 1 reflects some complicated insect physiology, and 4 presupposes a cultural environment and some human experience of psychology and sociodynamics. They are highly specific: example 1 tells about crickets, or more likely about one of the two thousand species of cricket, but nothing at all about any other insect, or about physiology in general.

There is a important distinction between these relationships and more fundamental relationships between physical quantities.

3.4   Dimensionless groups

Quantities with different dimensions can sometimes be combined to make dimensionless groups. For example:

$$
\begin{array}{ll}
\text{velocity } U & [LT^{-1}] \\
\text{gravitational acceleration } g & [LT^{-2}] \\
\text{length } a & [L]
\end{array}
$$

combine to form a group U^2/ga, which is dimensionless because

$$
\left[\frac{[LT^{-1}]^2}{[LT^{-2}][L]} \right] = \left[\frac{L^2 T^{-2}}{L^2 T^{-2}} \right] = [0]
$$

Sometimes we can find these groups by inspection, but we can also proceed systematically. Suppose we are looking for a dimensionless group of the form $U^\alpha g^\beta a^\gamma$. The dimension of $U^\alpha g^\beta a^\gamma$ is $[LT^{-1}]^\alpha [LT^{-2}]^\beta [L]^\gamma = [L^{\alpha+\beta+\gamma} T^{-\alpha-2\beta}]$.

If this is to be dimensionless, the power to which L appears must be 0, and so

$$
0 = \alpha + \beta + \gamma \tag{3.4.1}
$$

and the power to which T appears must be 0, and so

$$
0 = -\alpha - 2\beta \tag{3.4.2}
$$

Other fundamental dimensions M, θ, and Q do not appear at all, so that they do not give us any additional information.

Rearranging (3.4.2)

$$\beta = -\alpha/2 \tag{3.4.3}$$

and then substituting into (3.4.1)

$$\gamma = -\alpha/2 \tag{3.4.4}$$

and so the group $U^\alpha g^{-\alpha/2} a^{-\alpha/2} = (U/\sqrt{ga})^\alpha$ is dimensionless for any value of α.

We cannot necessarily form dimensionless groups out of any choice of quantities. For example, if we try to make a dimensionless group out of

> velocity U $[LT^{-1}]$
> gravitational acceleration g $[LT^{-2}]$
> density ρ $[ML^{-3}]$

the attempt fails, because M appears only in the dimension of ρ and cannot be balanced by anything else. If we proceed algebraically, and look for a dimensionless group of the form $Ug^\alpha \rho^\beta$, its dimensions are $[LT^{-1}][LT^{-2}]^\alpha[ML^{-3}]^\beta$ and the equations we get are

> for $\quad$ M $\quad 0 = \beta$
> for $\quad$ L $\quad 0 = 1 + \alpha - 3\beta$ $\tag{3.4.5}$
> for $\quad$ T $\quad 0 = -1 - 2\alpha$

Those three simultaneous equations in three variables have no solution. You can confirm that by adding together three times the first equation in (3.4.5), twice the second equation, and the third equation: the result is that 0 equals 1, which is not so. Alternatively, the first of (3.4.5) says that β is 0, and the third says that α is $-1/2$, but those values are not consistent with the second equation.

Engineers and scientists have found many dimensionless groups useful, and have often named them after people. $U/\sqrt{ga}$ is called Froude number, after William Froude (1810–1879) a naval architect who carried out fundamental work on ships (and built the first model test tank). It appears in many contexts of free surface flow of liquids, among them surface waves (such as waves on the sea), the resistance of ships, flow in open channels, and multiphase flow in pipelines and process plant.

Dimensionless groups are usually denoted by two-letter abbreviations, printed in roman type to distinguish them from symbols written in italic type. Froude number is printed Fr, whereas Fr denotes F multiplied by r. Generally, but not always, the simplest groups are the most useful ones.

3.5   Variables

Large parts of technology and science have to do with the relationships between quantities that can be measured, that have units, and that can take different values. Without making any prejudgements about what is or is not important, we might for example suppose that there is some relationship between the time it takes for a pendulum to swing back and forth, the length of the pendulum, the mass of the pendulum bob, the acceleration due to gravity, how far the pendulum swings, the temperature of the air the pendulum is swinging in, the color of the pendulum, and so on.

Similarly, there might be a relationship between how high a plant grows, the air temperature, the number of hours of daylight, how many hours of direct sun there are, how moist the soil is, how acid the soil is, how many kilograms of organic matter the soil contains within reach of the roots, and so on. How fast an airplane flies might be related to the thrust of its engines, the shape of the wings, the density of the air, and perhaps other quantities, such as the air viscosity and compressibility.

Dimensional analysis calls these different quantities variables. A large part of the art of scientific understanding of physical and biological systems is to learn which variables matter and which do not. That choice is often far from straightforward. Continuing the pendulum example, we might reason that:

— the *period* (the time it takes for the pendulum to swing back and forth) is significant, because it varies between pendulums;

— the *length* is significant, because a simple experiment with a piece of metal (such as a key) and a length of thread shows that a long pendulum swings more slowly than a short one;

- the *mass* of the pendulum bob might be significant, because the weight of the mass is the driving force;

- gravity plays a part, because without it the pendulum would not swing, and *gravitational acceleration* is the measure of the strength of gravity, though in an experiment on Earth we cannot vary gravity much;

- *how far the pendulum swings* might be important;

- the air *temperature* probably does not matter, because the behavior of a pendulum seems likely to be governed by the mechanical interaction between weight, mass, and length, and

- *the color* of the pendulum is almost certainly unimportant, because color has to do with light and the optical properties of the pendulum bob surface, and they are unlikely to interact with the mechanics.

It would be wise to keep an open mind about all these possibilities. Moreover, what parameters are important depends on what question we are trying to answer. A maker of a pendulum clock might have the priorities listed above, but the maker of a kinetic sculpture would have different ones: for him/her the color would be important, and the gravitational acceleration of little or no interest. The clockmaker's accountant would have different concerns again.

The task of deciding what is important is a difficult one, and dimensional analysis rarely provides a complete answer. Often we think of an idealization of the problem, in which we include some variables that appear significant — in the example, the period, length, mass, gravitational acceleration, and how far the pendulum swings — and tentatively exclude everything else. We can then explore the consequences of that idealization, and see if it gives sensible results that are consistent with other information, such as experiments or other kinds of theory.

Often it is useful to imagine an experiment in which we alter some of the variables, and examine the effect on other variables. In the pendulum example, we might alter the length, keeping everything else the same, and measure how the period varies. We can say that some of the quantities depend on some other quantities, and call the second group independent variables and the first group dependent variables. In the pendulum example, the period could be one dependent variable and the length one independent variable. But often the choice is arbitrary: if we had no familiarity with pendulums, and were observing them from the Moon, it might not be clear whether the period depends on the length or the length depends on the period, and for many purposes it does not matter which way round we interpret it.

The nomenclature is not ideal: a quantity might be treated as a variable even though there is no action we can take to vary it. Unless we go to the Moon or to another planet, we can only alter the gravitational acceleration by a very small amount. The velocity of light is believed to be absolutely fixed, and there is nothing at all we can do to vary it, but it can still be thought of as a variable.

3.6   Problems

Problems on dimensionless groups (Sec. 3.4)

3.1   A centrifugal pump is pumping water. The volume of water flowing through the pump in unit time is q, the speed of rotation of the pump is N revolutions in unit time, and the diameter of the pump rotor is D. The variables and their dimensions are

rotor diameter D $[L]$
volumetric flow rate q $[\text{volume/time}] = [L^3 T^{-1}]$
rotational speed N $[\text{revolutions/time}] = [[0]/[T]] = [T^{-1}]$

Form a dimensionless group out of these three variables.

3.2   Sound is transmitted through the air by a wave of compression and extension. The velocity of sound depends on the density of the air and on the compressibility, the fractional change in volume with respect to pressure. Taking the fundamental dimensions as M L and T, the variables and their dimensions are

sound velocity U $[LT^{-1}]$
air density ρ $[ML^{-3}]$
air compressibility β $[\text{fractional change of volume/change of pressure}]$

$$= \left[\frac{\text{volume/volume}}{\text{force/area}}\right] = \left[\frac{[L^3]/[L^3]}{[MLT^{-2}]/[L^2]}\right]$$
$$= [M^{-1}LT^2]$$

Form a dimensionless group out of these three variables.

3.3   Water flows past a cylindrical pile. Vortices leave the sides of the pile, alternately from one side and then the other. The frequency of vortex shedding depends on the velocity of the water and the diameter of the pile. The variables and their dimensions are

   pile diameter D [L]
   water velocity U $[LT^{-1}]$
   vortex frequency N [number of vortices/time] $= [[0]/[T]] = [T^{-1}]$

Form a dimensionless group out of these three variables.

3.4   A civil engineer is going to excavate a trench in clay. He/she wants to know how deep the trench can be, and still be stable and stand for a short time without support. The trench will be square in cross-section. The relevant variable describing the strength of the clay is a stress, a force per unit area, called by geotechnical engineers the undrained shear strength: undrained means that there is not enough time for water to move about within the clay. The clay density is important, and so is the gravitational acceleration, because if there were no gravity the clay would have no weight. The variables and their dimensions are

   trench depth H [L]
   clay density ρ $[ML^{-3}]$
   clay shear strength c [force/area] $= [MLT^{-2}/L^2] = [ML^{-1}T^{-2}]$
   gravitational acceleration g $[LT^{-2}]$

Form a dimensionless group out of these four variables. Show that you get the same result if you decide that you will take the fundamental dimensions as force [F] and length [L].

3.5   Surface tension is the property that allows a drop of water to hang at the end of a tap, that makes it possible carefully to float a needle on the surface of water, and that pulls a small droplet into a roughly spherical form. It is an energy per unit area, or equivalently a force per unit length. It is quite small, 0.073 N/m for water against air.

   One of the applications of the dimensionless group $U/\sqrt{ga}$ developed in Sec. 3.4 is to waves on the surface of water. If the waves are small ripples, the density ρ and the surface tension S are also significant. The variables are

length a	$[L]$
velocity U	$[LT^{-1}]$
gravitational acceleration g	$[LT^{-2}]$
water density ρ	$[ML^{-3}]$
water surface tension S	$[\text{force}/\text{length}] = [MLT^{-2}/L] = [MT^{-2}]$

Show that it is not possible to form a dimensionless group that includes ρ but not S, or to form a dimensionless group that includes S but not ρ. Form a group that includes both.

3.6 Another illustration of surface tension is seen by lowering the end of a drinking straw into a glass of water. The water rises in the straw, pulled up by the surface tension against the weight of the water. The smaller the diameter of the straw, the further the water rises. We want to find a relationship between the diameter of the straw D, the height of rise H, the surface tension S, the density ρ and the gravitational acceleration g. The variables are

diameter D	$[L]$
height of rise H	$[L]$
water surface tension S	$[\text{force}/\text{length}] = [MLT^{-2}/L] = [MT^{-2}]$
water density ρ	$[ML^{-3}]$
gravitational acceleration g	$[LT^{-2}]$

Capillary action is an important factor in the movement of water within soil. Construct a dimensionless group including the five variables.

Problems on dimensional consistency of equations (Sec. 3.3)

3.7 Which of the following equations are dimensionally consistent?

(a) $Q = 0.04V + W + 0.33A$

a formula used by heating contractors to determine the heating requirements of a room:

Q is the heat supply per °F temperature difference between inside and outside, in Btu/h °F,

V is the volume of the room, in ft^3

W is the area of the windows, in ft^2
A is the area of the external walls, in ft^2

(b) $T = 0.2Q/v$

the White formula for the tension left in a straight arc weld joining two steel plates (because of the cooling shrinkage of the weld metal)

T is the tension force
Q is the electrical power input to the welding arc
v is the speed at which the arc moves along the line of the weld

(c) $u = C\sqrt{\dfrac{AS}{L}}$

the Chezy formula for the mean flow velocity of water in a sloping pipe (whose cross-section is not necessarily circular)

u is the mean flow velocity
A is the cross-sectional area of the pipe
S is the slope of pipe (change of height/horizontal distance)
L is 'wetted perimeter' of cross-section
C is a constant

(d) $f = \frac{1}{2\pi}\sqrt{\dfrac{c^2 A}{VL}}$

which is the natural frequency of a Helmholtz acoustic resonator,[3] which has a short neck and then widens out into a large volume (like a beer bottle), and where

c is the speed of sound
A is the cross-sectional area of the neck
V is the volume
L is the length of the neck

If you have a musical ear, you can test this relationship by exciting the oscillations by blowing across the top of a beer bottle.

3.8   Here are four formulae for the lowest natural frequency f of a uniform flat plate vibrating in bending. The plate is a right-angled triangle with the shorter sides a and b; all three edges of the plate are clamped, so that they cannot move.

$$(\text{I})\ f = \frac{1}{2\pi}\sqrt{4004\frac{D}{mab}\left(\left(\frac{a}{b}\right)^2 + 1 + \left(\frac{b}{a}\right)^2\right)}$$

where a is the length of one of the shorter sides, b the length of the other shorter side, m the mass per unit area of the plate, t the thickness of the plate, D is $Et^3/12(1 - v)$, E the Young's modulus, dimensions [stress] = [force/area], and v is the Poisson's ratio (dimensionless).

$$(\text{II})\ f = \frac{1}{2\pi}\sqrt{4004\frac{D}{ma^2b^2}\left(\left(\frac{a}{b}\right)^2 + 1 + \left(\frac{b}{a}\right)^2\right)}$$

$$(\text{III})\ f = \frac{1}{2\pi}4004\frac{D}{mab}\left(\left(\frac{a}{b}\right)^2 + 1 + \left(\frac{b}{a}\right)^2\right)$$

$$(\text{IV})\ f = \frac{1}{2\pi}\sqrt{4004\frac{D}{ma^2b^2}\left(\left(\frac{a}{b}\right)^2 + 1 + 3\left(\frac{b}{a}\right)^2\right)}$$

Show by dimensional analysis and symmetry arguments that three of the four formulae must be incorrect. Which three?

The correct formula was used in the analysis of the effects of the 1988 explosion on the Piper Alpha oil production platform.

Problems on variables (Sec. 3.5)

Each of the following describes a situation taken from technology and science. You are asked to make a tentative decision about which physical quantities are likely to be of primary importance, which are likely to be of secondary importance, and which are probably not important at all. Be aware that there is no one 'right' answer. Some of the situations are very complex, and even supposedly knowledgeable 'experts' may well disagree. Everyday experience is relevant. If you are uncertain, do not be discouraged.

In the first three examples, some of the possible relevant quantities are listed.

3.9   An airplane is powered by two jet engines. You wish to determine the thrust required from each jet to drive the airplane at a given speed. Relevant physical quantities might be:

- the linear dimensions of the airplane
- the density of the air
- the viscosity of the air
- the mass of the airplane
- the positions of the control surfaces (ailerons, rudder, flaps, etc.)
- the compressibility of the air
- for the air, the ratio between specific heat at constant pressure and specific heat at constant volume
- the number of crew and passengers

3.11   A fire breaks out in an oil terminal, as happened at the Buncefield depot in England in December 2005. A plume of smoke rises into the sky. Air traffic controllers wish to know the heights above ground of the top and bottom of the smoke. Relevant physical quantities might be:

- the density of the air outside the smoke plume
- the temperature of the air outside the smoke plume
- the rate at which the fire releases energy
- the specific heat of the air
- the temperature at the heart of the fire
- the horizontal distance between the fire and the location at which the smoke heights are wanted
- the velocity of the wind at different heights
- the composition of the oils stored in the terminal.

3.12   Waves break against a breakwater built of lumps of rock. You wish to determine how large the rocks have to be if they are to be stable and remain in position. Relevant physical quantities might be:

- the density of the water
- the density of the rock
- the speed of the wind
- the grain size of the seabed
- the height of the waves
- the period of the waves
- the strength of the rock
- the density of the air
- the fracture toughness of the rock
- the color of the rock
- the shapes of the lumps of rock
- the age of the rock
- whether or not the waves are breaking before they reach the breakwater.

3.13 A well with a hand-driven pump is going to supply water to a village in Africa. The villagers want to know how much water it will supply.

3.14 An earth dam begins to leak. River engineers want to know how much water will flow into the river downstream.

3.15 An earth dam begins to leak, and the water flowing through the dam begins to carry with it particles of soil, so that the leak enlarges. River engineers want to know how long it will be before the dam collapses.

3.16 Mountain villagers who have lost their houses in an earthquake are supplied with insulated tents. The winter is coming, and a relief agency wants to know what the temperature will be inside the tents, and how much ventilation needs to be provided to keep the carbon dioxide content of the air inside to 1000 ppm.

3.17 An oil pipeline splits open close to the bank of a river, and some time passes before the flow can be stopped. The authorities want to know how much oil will flow into the river.

3.18   Paint is applied with a brush. The painter wishes to know how thick the paint will be.

3.19   The paint in 3.18 is water-based, so that it dries by evaporation. The painter wishes to know how long it will be before the paint is dry.

3.20   Paint is applied to a wall with an air-driven spray. The painter wishes to know how thick the layer of paint will be.

3.21   A leg of lamb is cooked in an oven. The cook wishes to know the cooking time required for the temperature at the center of the leg to reach 80 °C, the temperature needed to destroy tapeworm eggs. How would the answer be different for a fan oven?

3.22   A glassblower forms glass when it is a thick viscous liquid. He/she holds a hollow glob of molten glass on a blowpipe, expands it into a larger bubble, forms it into a round cup, and then expands the cup into a circular plate by rotating the blowpipe rapidly. What variables are important at different stages of this process?

3.23   Plate glass is made by floating molten glass onto molten tin. What factors will determine how thick the glass is and how flat it is?

References

1. McAllister, EW (ed.) (1998). *Pipeline Rules of Thumb Handbook.* Houston, TX: Gulf Publishing.

2. Pippard, AB (1972). *Forces and Particles: An Outline of the Principles of Classical Physics.* Macmillan.

3. Dowling, AP and JE Ffowcs-Williams (1983). *Sound and Sources of Sound.* New York: Ellis Horwood.

Chapter 4

Dimensional Analysis

4.1 Relations between dimensionless groups

Fundamental relationships are dimensionally consistent and unit-free. The central idea of dimensional analysis is that those relationships can always be expressed as relationships between dimensionless groups.

This is the Vaschy–Buckingham Pi theorem. Pi stands for the Greek capital letter Π for product, not the ratio π between a circle's circumference and its diameter. The idea can be demonstrated by an example. We start with a general relationship expressed in terms of variables that have units, and show that by appropriate changes of units we arrive at a relationship expressed in terms of dimensionless groups.

First, consider a very simple mechanical system. People interested in mechanics have always been fascinated by pendulums. Pendulums exist in nature: fruit hangs from branches, and monkeys swing on creepers from one tree to the next. The first human observation of a pendulum must have happened far into prehistory. Some four million years ago a family of three left their footprints in the Olduvai Gorge in Kenya. We can imagine that the family paused one evening, and that one of them tied a stone to a length of animal sinew, held the other end, watched the stone swing, and wondered why it swung. She/he might have noticed that how long it takes to swing does not seem to depend much on the whether the swing is large or small, that how long it takes seems not to depend on how large the stone is, and that it is the same day after day and hour after hour. She or he might also have seen that a longer pendulum swings more slowly than a shorter one. None of these things are immediately obvious from just thinking about a pendulum.

Dimensional analysis cannot tell us everything about pendulums, but it takes us surprisingly far.

A simple pendulum consists of a concentrated mass (bob) connected to an immovable support by a string. The string has no mass and is inextensible, so that it does not stretch in response to changing tensions in it. The pendulum swings in a vacuum, so that air resistance and the inertia of the surrounding air do not come into the problem. One of the pendulum's properties is the period of oscillation, the time it takes to swing from one momentarily stationary position to the other and then to swing back to the starting point. We wish to find how the period of oscillation, denoted t, depends on the other variables.

The period t might depend on:

— the length s
— the gravitational acceleration g
— the mass m
— the amplitude a (a length measured horizontal from one extreme of the oscillation to the other)

There is no obvious reason to exclude any of these variables. The length seems likely to be important, and a simple observation indicates that a long pendulum swings more slowly than a short one. Acceleration due to gravity comes in, because if there were no gravity the pendulum would not swing at all. Mass comes in, because the weight of the mass is the driving force.

Five quantities are involved, and their dimensions are:

period t	$[T]$
length s	$[L]$
gravitational acceleration g	$[LT^{-2}]$
mass m	$[M]$
amplitude a	$[L]$

choosing the fundamental dimensions as M, L, and T. We have no need for temperature and electric charge in this problem.

The period t is some function of the other four variables $s, g, m,$ and a, measured in some units, which we call 'old' units. We can write the functional relationship as

$$t = f(s, g, m, a) \tag{4.1.1}$$

So far we know nothing at all about the form of the function f. We now change the units, in just the same way as we changed units in specific instances in Chap. 2, and let

1 new unit of mass $= \mu$ old units of mass
1 new unit of length $= \lambda$ old units of length
1 new unit of time $= \tau$ old units of time

Substituting into (4.1.1), in the new units it becomes

$$\frac{t}{\tau} = f\left(\frac{s}{\lambda}, \frac{g}{\lambda/\tau^2}, \frac{m}{\mu}, \frac{a}{\lambda}\right) \tag{4.1.2}$$

which is true for any values of the three conversion factors $\mu, \lambda,$ and τ. Now we give the three factors specific values, and let

$$\mu = m$$
$$\lambda = s$$
$$\tau = \sqrt{\frac{s}{g}} \tag{4.1.3}$$

and (4.1.3) becomes

$$\frac{t}{\sqrt{\frac{s}{g}}} = f\left(1, 1, 1, \frac{a}{s}\right) = f^*\left(\frac{a}{s}\right) \tag{4.1.4}$$

where f^* is another function, but one whose form is still unknown.

The original relationship (4.1.1) had five variables $t, s, g, m,$ and a, and all five of them had dimensions. Now (4.1.1) has been transformed into a simpler relationship (4.1.4) between just two dimensionless groups, $t/\sqrt{s/g}$ and a/s. Rearranging (4.1.4)

$$t = \sqrt{\frac{s}{g}} f^*\left(\frac{a}{s}\right) \tag{4.1.5}$$

so that if the amplitude ratio a/s is constant

- t is proportional to $\sqrt{s}$
- t is inversely proportional to $\sqrt{g}$
- t is independent of m.

or, in words,

- the period is proportional to the square root of the length of the pendulum,
- the period is inversely proportional to the square root of the gravitational acceleration,
- the period does not depend on the mass of the pendulum bob.

All three statements are nontrivial. They are not by any means obvious. We have not had to go into the underlying mechanics of a pendulum, and have simply exploited the notion elegantly stated by Richardson[1]

> "...that phenomena go their way independently of the units whereby we measure them."

The independence of m occurs because — though we started by assuming that m might be involved — the process of changing the units and selecting the scaling factor eliminated m. If we think about why it might happen, it is because the force exerted by gravity on a mass is proportional to the mass, but Newton's law says that acceleration is proportional to force divided by mass, and therefore, if force is proportional to mass, acceleration must be independent of mass (just as the acceleration of a body falling in a vacuum is independent of its mass). However, we can confidently make use of (4.1.5) without going through that reasoning, and without knowing anything about Newton's law.

That is as far as dimensional analysis can take us. To go further, we either have to do some experiments or some more serious theory.

If we had no prior knowledge of pendulums, and needed to be able to predict the period from the four variables, we could in principle vary all four variables, measure the period corresponding to each combination, arrive at a large amount of data, and try to puzzle out the relationship.

A further complication is that it is difficult to vary g: the variations on the surface of the earth are small, about 1 part in 1000, and to achieve big variations we would have to put the experiment in a spaceship or take it to the moon.

Equation (4.1.4) shows that that laborious and difficult process can be avoided. All we need to do is to vary a/s, measure $t/\sqrt{(s/g)}$, and plot a relationship between two variables, a/s and $t/\sqrt{(s/g)}$. Dimensional analysis has hugely reduced the experimental and analytical effort involved, and enabled us to express the results in a very simple form.

In fact it turns out that if a/s is small f^* is independent of a/s and equal to 2π, but to show that requires different arguments or an experiment: see Problem 4.1.

4.2   Proof of Pi theorem

Aimé Vaschy (1857–1899)[2] and much later Buckingham[3] (1867–1940) showed that fundamental relationships can always be expressed as relationships between dimensionless groups. In English-speaking countries this 'Pi theorem' is attributed to Buckingham, doubtless in part an example of Anglo-Saxon provincialism, but perhaps also because Vaschy's paper proved the result in fewer than two elegant but terse pages and was thought somewhat obscure. Vaschy's research is much better known in other countries. Vaschy did not use the word dimension, and did not say precisely how many groups there would be. Buckingham clarified the question of the number of groups, and was extremely good at publicizing dimensional analysis. Sterrett[4] has a fascinating discussion of Buckingham and the relation between his work and the philosopher Wittgenstein.

The reader who is ready to accept the results and wants to move on to applications may skip this section, and jump to Sec. 4.3.

Vaschy's proof is a generalization of the approach taken in the pendulum example of Sec. 4.1, where the fact that the fundamental units are arbitrary allowed us to choose them in a special way that reduced a relationship between five variables to a relationship between two dimensionless groups. A translation of his proof follows.

"The most general law of similarity in mechanics and physics results from the following theorem:

Let $a_1, a_2, a_3, \ldots, a_n$ be physical quantities, of which the first p are distinct fundamental units and the last $(n - p)$ are derived from the p fundamental units (for example, a_1 could be a length, a_2 a mass, a_3 a time, and the $(n - 3)$ other quantities would be forces, velocities, etc.: then $p = 3$). If between these n quantities there exists a relation

$$F(a_1, a_2, \ldots, a_n) = 0 \qquad (4.2.1)$$

that remains the same whatever the arbitrary magnitudes of the fundamental units, this relationship can be transformed into another relationship between at most $(n - p)$ variables, that is

$$f(x_1, x_2, \ldots, x_{n-p}) = 0 \qquad (4.2.2)$$

the variables $x_1, x_2, \ldots, x_{n-p}$ being monomial functions of $a_1, a_2, \ldots, a_n$ (for example $x_1 = A a_1^{\alpha 1} a_2^{\alpha 2} \cdots a_n^{\alpha n}$).

In the special case where $n = 5$ and $p = 3$, one recovers the theorem given by M. Carvallo.

To demonstrate the theorem that we have just stated, we note that since the quantities $a_{p+1}, a_{p+2}, \ldots, a_n$ correspond to derived units, that is the same as saying that one can find exponents $\alpha, \beta, \ldots, \alpha', \beta'$ in such a way that the ratios

$$\frac{a_{p+1}}{a_1^{\alpha} a_2^{\beta} \cdots a_p^{\lambda}} = x_1, \qquad \frac{a_{p+2}}{a_1^{\alpha'} a_2^{\beta'} \cdots a_p^{\lambda'}} = x_2, \ldots \qquad (4.2.3)$$

are independent of the arbitrary values of the fundamental units. (Thus, if a_1, a_2, a_3, and a_4 were to denote a length, a mass, a time, and a force respectively, the ratio $\frac{a_4}{a_1 a_2 a_3^{-2}}$ would have a value independent of the choice of units). Now, the given relation (4.2.1)

$$F(a_1, a_2 \cdots a_p, a_{p+1}, a_{p+2} \cdots) = 0 \qquad (4.2.4)$$

can be written

$$F(a_1, a_2 \cdots a_p, x_1 a_1^\alpha \cdots a_p^\lambda, x_2 a_1^{\alpha^r} \cdots a_p^{\lambda^r} \cdots) = 0 \qquad (4.2.5)$$

or more simply

$$f(a_1, a_2 \cdots a_p, x_1, x_2 \cdots x_{n-p}) = 0 \qquad (4.2.6)$$

But, by causing the magnitudes of the fundamental units to vary, one can vary arbitrarily the numerical values of the quantities $a_1, a_2, \ldots, a_p$ whose intrinsic magnitudes are considered fixed, while the numerical values of $x_1, x_2, \ldots, x_{n-p}$ will not change at all. Since the relationship (4.2.6) must remain the same, whatever the arbitrary values of $a_1, a_2, \ldots, a_p$, it must be independent of those parameters; that relation therefore takes the simplest form (4.2.2)." ■

Vaschy goes on to consider two examples. The first is the period of a pendulum, analyzed in much the same way as here in Sec. 4.1. The second is the current at one end of a telegraph cable, as a function of time since a battery was connected to the other end. He points out that in electrical problems there are four distinct fundamental units, and for the purpose of his problem takes them to be time, length, electric capacity, and energy. This again emphasizes that there is some freedom of choice of fundamental units.

Vaschy says that the number of independent dimensionless groups is 'at most $n - p$,' which leaves open the question of exactly how many groups there are. He also implicitly assumes that the first p variables $a_1, a_2, a_3, \ldots, a_p$ have the dimensions of the fundamental units. The second assumption can be easily removed by choosing to define the fundamental units as having the dimensions of the first p variables. Buckingham[3] clarifies the first question.

Consider the general problem of finding a dimensionless group of the form $a_1^{\alpha 1} a_2^{\alpha 2} \cdots a_n^{\alpha n}$ in a problem with n variables and p fundamental dimensions. Each dimensionless group corresponds to a choice of the n exponents $\alpha 1, \alpha 2, \ldots, \alpha n$. Each independent group corresponds to an independent choice of the exponents. Following Vaschy — but making a small change in notation to avoid subscripts on exponents — if $a_1^{\alpha 1} a_2^{\alpha 2} \cdots a_n^{\alpha n}$ is dimensionless, then

$$[0] = [a_1^{\alpha 1}\ a_2^{\alpha 2}\ \cdots\ a_n^{\alpha n}]$$
$$= [\text{dimension of } a_1]^{\alpha 1}\ [\text{dimension of } a_2]^{\alpha 2} \cdots [\text{dimension of } a_n]^{\alpha n}$$

$$(4.2.7)$$

If this condition is met for the first of the fundamental dimensions

$$0 = d_{1,1}\alpha 1 + d_{1,2}\alpha 2 + \cdots + d_{1,n}\alpha n \qquad (4.2.8)$$

where $d_{i,j}$ denotes the component of the jth variable a_j that corresponds to the ith fundamental dimension. For example, if there are three fundamental dimensions M, L, and T (taken in that order), and the first variable a_1 is a force whose dimension is $[\text{MLT}^{-2}]$, then $d_{1,1}$ is $+1$, $d_{2,1}$ is $+1$, and $d_{2,1}$ is -2. If the second variable a_2 is a density whose dimension is $[\text{ML}^{-3}]$, then $d_{1,2}$ is 1, $d_{2,2}$ is -3, and $d_{2,1}$ is 0.

If the conditions of (4.2.7) are met for the second of the fundamental dimensions

$$0 = d_{2,1}\alpha 1 + d_{2,2}\alpha 2 + \cdots + d_{2,n}\alpha n \qquad (4.2.9)$$

and so on for each of the fundamental dimensions, down to

$$0 = d_{p,1}\alpha 1 + d_{p,2}\alpha 2 + \cdots + d_{p,n}\alpha n \qquad (4.2.10)$$

for the pth dimension. Putting all p equations together, the condition for $a_1^{\alpha 1} a_2^{\alpha 2} \cdots a_n^{\alpha n}$ to be a dimensionless group is that the n exponents $\alpha 1, \alpha 2, \ldots, \alpha n$ form a solution of the p simultaneous equations

$$
\begin{aligned}
0 &= d_{1,1}\alpha 1 + d_{1,2}\alpha 2 + \cdots + d_{1,n}\alpha n \\
0 &= d_{2,1}\alpha 1 + d_{2,2}\alpha 2 + \cdots + d_{2,n}\alpha n \\
&\ \vdots \\
0 &= d_{p,1}\alpha 1 + d_{p,2}\alpha 2 + \cdots + d_{p,n}\alpha n
\end{aligned}
\qquad (4.2.11)
$$

The equations can be more conveniently written in matrix–vector notation, as

$$0 = \mathbf{D}\alpha \qquad (4.2.12)$$

where $\mathbf{0}$ is the vector $[0 \cdots 0]$ with p components, all of them zero; $\mathbf{D}$ is the matrix of dimensions of the variables, with p rows and n columns:

$$\begin{bmatrix} d_{1,1} & d_{1,2} & \cdots & d_{1,n} \\ d_{2,1} & d_{2,2} & \cdots & d_{2,p} \\ \cdot & \cdot & \cdots & \cdot \\ d_{p,1} & \cdot & \cdots & d_{p,n} \end{bmatrix} ; \quad \text{and}$$

α is the vector $[\alpha1\ \alpha2\ \alpha3\ \cdots\ \alpha n]$ with n components.

How many independent solutions there are depends on the matrix $\mathbf{D}$. This theme is extensively discussed in texts on linear algebra: see, for example, Strang.[5] If n is equal to p, $\mathbf{D}$ is square, and (4.2.12) generally has no solution, except in the special case when $\mathbf{D}$ is singular, that is, when its determinant $|\mathbf{D}|$ is 0. Except in that case, no dimensionless group can be found.

If n is greater than p, (4.2.12) generally has $n - p$ independent solutions. An exception occurs when one or more of the rows of $\mathbf{D}$ is a linear combination of two or more of the other rows, in mathematical terms, when the rank r of $\mathbf{D}$ is less than p. In that case the number of independent solutions is $n - r$ rather than $n - p$.

The number of independent dimensionless groups is therefore $n-p$ unless r is less than p, in which case it is $n - r$.

4.3   Revisiting the example of Sec. 4.2, starting from the Pi theorem

It is often more rapid and convenient to start with the notion of dimensionless groups, rather than by transforming the variables in the way illustrated in Sec. 4.1.

Consider again the simple pendulum, and once more begin by listing the different quantities and their dimensions:

period t	[T]
length s	[L]
gravitational acceleration g	$[LT^{-2}]$

$$\text{mass } m \qquad [M]$$
$$\text{amplitude } a \qquad [L]$$

choosing the fundamental dimensions as M, L, and T.

There are five quantities, and three fundamental dimensions, and so we expect a relationship between $5 - 3 = 2$ independent dimensionless groups.

We start by looking for a dimensionless group composed from $t, s, g,$ and m, leaving a on one side for the moment. Look for a dimensionless group of the form:

$$tm^{\alpha}s^{\beta}g^{\gamma}$$

whose dimensions are

$$[T][M]^{\alpha}[L]^{\beta}[LT^{-2}]^{\gamma}$$

and so

$$
\begin{array}{ll}
M & 0 = \alpha \\
L & 0 = \beta + \gamma \\
T & 0 = 1 - 2\gamma
\end{array}
\qquad (4.3.1)
$$

The solution is

$$
\begin{array}{l}
\alpha = 0 \\
\beta = -1/2 \\
\gamma = +1/2
\end{array}
\qquad (4.3.2)
$$

and so $t\sqrt{(g/s)}$ is dimensionless.

Amplitude a has not yet been considered. The ratio a/s is a second independent dimensionless group. The two groups are $t\sqrt{(g/s)}$ and a/s. There are many other dimensionless groups that can be formed out of the five quantities, but no other *independent* groups. Other dimensionless groups are combinations of those two: for instance, $tg^{1/2}s^{3/2}/a^2$ is dimensionless, but is $\{t\sqrt{(g/s)}\}/\{a/s\}^2$. Any dimensionless group that only involves $t, g,$ and s is a power of $t\sqrt{(g/s)}$: for instance, s/gt^2 is dimensionless, but is $(t\sqrt{(g/s)})^{-2}$.

We now have two dimensionless groups, $t\sqrt{(g/s)}$ and a/s, composed from four variables t, g, s, and a. No other independent groups can be made from the same four variables. It follows from the Pi theorem that $t\sqrt{(g/s)}$ must be a function of a/s, so that

$$t\sqrt{\frac{g}{s}} = F\left(\frac{a}{s}\right) \tag{4.3.3}$$

where F is unknown, which is the result secured by a different route in Sec. 4.1. Multiplying by $\sqrt{(s/g)}$

$$t = \sqrt{\frac{s}{g}} F\left(\frac{a}{s}\right) \tag{4.3.4}$$

as before.

A skeptical reader might at this point wonder if the results somehow depend on the choice of groups or on the choice of fundamental dimensions. They do not, and to convince ourselves we explore what happens if we form the groups differently, or if we chose different fundamental dimensions.

4.4 Choosing the dimensionless groups and fundamental dimensions in a different way

If we chose to make the first group out of t, m, a, and g rather than t, m, s, and g, we find

$$t\sqrt{\frac{g}{a}} = F^*\left(\frac{a}{s}\right) \tag{4.4.1}$$

with a different function F^*, so that if the amplitude ratio a/s is constant

- t is proportional to $\sqrt{a}$ (and therefore to $\sqrt{s}$ since a/s is constant)
- t is inversely proportional to $\sqrt{g}$
- t is independent of m

which are the same results as before. This alternative formulation is just as correct as the first one. In fact it is not quite so convenient, because F^*

turns out to depend much more strongly on a/s than does F in Eq. (4.3.4), but that again depends on information beyond dimensional analysis.

Instead of choosing M, L, and T as fundamental dimensions, we could choose mass M, length L, and force F. What are the dimensions of time in these terms? [Acceleration] is force divided by mass, and with these alternative fundamental dimensions the dimension of acceleration is

$[M^{-1}F]$. Acceleration is also (length)/(time)^2, and so

$$[\text{time}] = \left[\sqrt{\left(\frac{\text{length}}{\text{acceleration}}\right)}\right] = \sqrt{\left[\frac{L}{M^{-1}F}\right]} = [M^{1/2}L^{1/2}F^{-1/2}]$$

and the dimensions of the different quantities become

period t	$[M^{1/2}L^{1/2}F^{-1/2}]$
length s	$[L]$
gravitational acceleration g	$[M^{-1}F]$
mass m	$[M]$
amplitude a	$[L]$

Forming a dimensionless group from t, m, s, and g, if $tm^\alpha s^\beta g^\gamma$ is dimensionless, then

$$[M] \quad 0 = +\frac{1}{2} + \alpha - \gamma$$

$$[L] \quad 0 = +\frac{1}{2} + \beta \tag{4.4.2}$$

$$[F] \quad 0 = -\frac{1}{2} - \gamma$$

whose solution is

$$\alpha = 0$$

$$\beta = -\frac{1}{2} \tag{4.4.3}$$

$$\gamma = +\frac{1}{2}$$

so that the dimensionless group is $t\sqrt{(g/s)}$. The second group is a/s. The same results as before all follow.

This shows that the conclusions do not depend on the choice of underlying fundamental dimensions.

4.5   Resistance of ships

The force required to move a ship through the sea has two principal components. The ship has to push the water aside, the sideways movement implies acceleration of the water, the water has mass, and so the ship has to exert forces on the water. The motion of the ship creates waves, waves require force to generate them (because the water has mass), and the wave energy is radiated away from the ship. The forces between the water and the ship generate a resistance to motion. Exactly the same thing happens if you move your hand rapidly through water in a swimming pool. The other component is viscous resistance; the motion deforms the water, and energy is dissipated in overcoming the water's viscous resistance to being deformed rapidly: this is the resistance you feel if you move a spoon slowly through a pot of honey.

The following analysis is initially concerned with the first component of resistance, which is related to the inertia of the water and is the dominant component for high-speed ships. We want to find how the force D required to move the ship at a steady speed U depends on other variables. The geometry of a ship is plainly complicated, and a full description would require lots of dimensions, but we consider a series of geometrically similar ships, each one a scaled version of the others, so that the dimensions of a ship are completely characterized by its length.

We can expect the resistance, a force

$$\text{resistance } D \qquad\qquad [MLT^{-2}]$$

to depend on:

$$
\begin{array}{ll}
\text{the speed } U & [LT^{-1}] \\
\text{the ship's length } s & [L] \\
\text{the density of the water } \rho & [ML^{-3}] \\
\text{the gravitational acceleration } g & [LT^{-2}]
\end{array}
$$

Density comes in because if the water had no density it would have no inertia, and there would be no inertial resistance to the water being moved aside by the ship. Gravitational acceleration comes in because if there were no gravity the water would stay heaped up by the ship passing through it, and there would be no waves. Common experience suggests that velocity and size are both significant. If you move you hand fast through water, you feel more resistance than if you move your hand slowly, and if you move your finger through the water you feel less resistance than if you move your whole hand at the same speed. Similarly, a ship moving fast will encounter more resistance than the same ship moving slowly, and a large ship will meet a higher resistance than a smaller geometrically similar ship.

Again there are five quantities and three dimensions, and $5 - 3 = 2$ dimensionless groups. Look for one group containing D, U, s, and ρ, of the form

$$DU^{\alpha}s^{\beta}\rho^{\gamma}$$

whose dimensions are

$$[MLT^{-2}][LT^{-1}]^{\alpha}[L]^{\beta}[ML^{-3}]^{\gamma}$$

If the group is dimensionless

$$
\begin{array}{ll}
M & 0 = 1 + \gamma \\
L & 0 = 1 + \alpha + \beta - 3\gamma \\
T & 0 = -2 - \alpha
\end{array}
\qquad (4.5.1)
$$

and so

$$
\begin{array}{l}
\alpha = -2 \\
\beta = -2 \\
\gamma = -1
\end{array}
\qquad (4.5.2)
$$

and the required group is $D/\rho U^{2}s^{2}$.

So far we have made a group out of D, ρ, U and s, but g is not involved. We need a second independent dimensionless group to bring in g. The simplest group of that type is $U/\sqrt{(gs)}$, the Froude number Fr derived in Sec. 3.5. Now we have two dimensionless groups, and between them the groups include all five variables D, ρ, U, s, and g. One group must be a function of the other group. Accordingly

$$\frac{D}{\rho U^2 s^2} \quad \text{is a function of} \quad \frac{U}{\sqrt{(gs)}}$$

so that the dimensionless resistance $D/\rho U^2 s^2$ is a function of Fr.

This can be exploited to predict the resistance of a ship from a test on a model. If we construct a model ship so that is geometrically similar to the full-scale ship, and tow the model in a towing tank at a speed chosen to that $U/\sqrt{(gs)}$ is the same in the model as in the prototype, then $D/\rho U^2 s^2$ must be the same in the model as in the prototype. If we measure D on the model, we can calculate D for the full-scale ship. Towing tests of this kind are widely used, because a ship has a complicated shape and until recently the calculation of resistance was far beyond the reach of computational fluid dynamics. Figure 4.1 shows a model yacht in a towing tank and Fig. 4.2 a model frigate in waves, both at the Institute for Ocean Technology in St. John's in Canada.

For example, consider a prototype ship 150 m long with a prototype speed of 20 knots (10.3 m/s). A 1/50 scale model is $150/50 = 3$ m long, which is convenient for a towing tank. Since $U/\sqrt{(gs)}$ is the same for model and prototype

$$\frac{U_m}{\sqrt{g_m s_m}} = \frac{U_p}{\sqrt{g_p s_p}} \tag{4.5.3}$$

where subscript m stands for model and p for prototype, and so

$$\frac{U_m}{U_p} = \sqrt{\frac{g_m}{g_p}} \sqrt{\frac{s_m}{s_p}} = \sqrt{1}\sqrt{\frac{1}{50}} = \frac{1}{\sqrt{50}} \tag{4.5.4}$$

since g is the same in the tank as in the sea. Then

$$U_m = 20\,\text{knots} \times \frac{1}{\sqrt{50}} = 2.83\,\text{knots} \tag{4.5.5}$$

Fig. 4.1. Model yacht in towing tank (by courtesy of the Institute for Ocean Technology).

Fig. 4.2. Model frigate in wave tank (by courtesy of the Institute for Ocean Technology).

which is 1.46 m/s (5.24 km/h), a brisk walking speed, and in that case

$$\frac{D_m}{\rho_m U_m^2 s_m^2} = \frac{D_p}{\rho_p U_p^2 s_p^2} \tag{4.5.6}$$

If the measured D_m for the model in the tank is 18 N, and the tank contains fresh water (density 1000 kg/m^3), the resistance D_p of the full-scale ship at 20 knots in the sea (density 1025 kg/m^3) is given by rearranging (4.5.6) into

$$\frac{D_p}{D_m} = \left(\frac{\rho_p}{\rho_m}\right)\left(\frac{U_p}{U_m}\right)^2\left(\frac{s_p}{s_m}\right)^2 = \left(\frac{1025}{1000}\right)\left(\sqrt{50}\right)^2 (50)^2 = 1.281 \times 10^5$$

and so

$$D_p = (1.281 \times 10^5)(18\,\text{N}) = 2.31\,\text{MN}$$

The power required to drive the ship is 2.31 MN $\times$ 10.3 m/s = 23.8 MW (31,900 HP).

This is not the whole story about the resistance of ships, because it only deals with wave-making resistance, related to gravity and the inertia of the water the ship is floating in. It implicitly assumes that the fluid has no resistance to shear deformation. This does not account for the different viscosities of different fluids: honey, water, tar, and petrol have rather similar densities, but honey and tar require more force to deform them than water and petrol do. We can expect that a ship sailing in a sea of honey or tar would have a higher resistance to motion than the same ship sailing in water or petrol. That suggests that another material variable besides density might be involved (if you already know about viscosity, you can skip forward).

We can use dimensional analysis to deduce the dimensions of that variable. Imagine two horizontal parallel plates, each of area A, separated by a small distance h, and the space between the plates filled with fluid (Fig. 4.3a). The lower plate is held stationary, and the upper plate is moved at a uniform velocity U parallel to the lower plate. A force F is required. The force depends on the unknown variable, the relative velocity U, the area A, and the separation h. Look for a relationship of the form

$$F = (\text{parameter})\, U^\alpha A^\beta h^\gamma \tag{4.5.7}$$

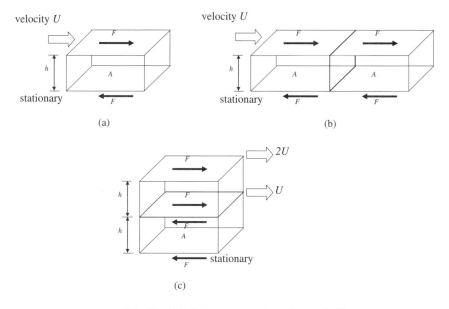

Fig. 4.3. *Parallel plates separated by a viscous fluid.*

Imagine two such systems put side by side, both with the same U and the same h (Fig. 4.3b). The total area has increased from A to $2A$, and the total force from F to $2F$. This indicates that β must be 1.

Now imagine one system put on top of a second one (Fig. 4.2c). The relative velocity between the lowest plate and the middle one is U, and the relative velocity between the middle plate and the uppermost one is U. The middle plate has a force in the negative U direction exerted on it by the fluid below, and a force in the positive U direction exerted by the fluid above. If we took away the middle plate, we would not alter the flow: we would then have a system with area A, relative velocity $2U$ between the lowest plate and the upper most, and separation $2h$ between the lowest plate and the uppermost, but still the same force F. Accordingly, if we increase h and U by the same multiplier, F is unchanged. In terms of the relationship (4.5.7) in algebraic terms, $\alpha + \gamma = 0$.

Finally, we find experimentally that in many fluids (but not all), the force is proportional to the velocity U, so that $\alpha = 1$. Fluids that have this property are called Newtonian or linear viscous. Gases, water, and many kinds of oil behave as Newtonian fluids. Blood, some kinds of heavy crude oil, paint, and tomato ketchup have more complicated non-Newtonian behavior, and will not be further considered here.

Putting these results together

$$\alpha = 1$$
$$\beta = 1 \qquad\qquad (4.5.8)$$
$$\gamma = -1$$

and so

$$F = (\text{parameter})\frac{UA}{h} \qquad\qquad (4.5.9)$$

and the parameter we are seeking has the same dimensions as $[hF/UA]$, which is

$$\frac{[L][MLT^{-2}]}{[LT^{-1}][L^2]} = [ML^{-1}T^{-1}] \qquad\qquad (4.5.10)$$

Fluid mechanics calls this parameter *viscosity*, and denotes it μ. Sometimes it is called *dynamic viscosity* to distinguish it from another kind of viscosity, unfortunately called *kinematic viscosity*. The dimension of viscosity is the dimension of pressure $[ML^{-1}T^{-2}]$ multiplied by time $[T]$. The SI unit of viscosity is the Pascal second (Pa s).

Continue the problem of resistance of ships, but now generalize it to add viscosity to the relevant properties of the fluid the ship is sailing through. The resistance

$$D \qquad\qquad\qquad [MLT^{-2}]$$

now depends on:

the speed U $\qquad\qquad [LT^{-1}]$
the ship's length s $\qquad\qquad [L]$
the density of the water ρ $\qquad [ML^{-3}]$
the gravitational acceleration g $\quad [LT^{-2}]$
the viscosity of the water μ $\qquad [ML^{-1}T^{-1}]$

The dimensionless resistance $D/\rho U^2 s^2$ and the Froude number $U/\sqrt{(gs)}$ were useful before, and there is no reason to give them up. We look for a third dimensionless group that brings in the viscosity. It must include either D or ρ, because their dimensions are the only ones beside μ that include M. We do not want to include D in the new group, if we can avoid doing so, because then we would have two groups that both include the resistance D, which is what we want to predict. The new group ought therefore not to include D. If the new group includes μ it must also include ρ, and μ and ρ must appear as the ratio μ/ρ, since otherwise the group cannot be dimensionless. The dimension of μ/ρ is $[L^2 T^{-1}]$.

There are still some choices. Ought the rest of the new group to include U, s, or g? The acceleration g is relevant to waves, but a ship sailing very slowly in a sea of honey would not make waves, and its resistance would depend on viscosity but not gravity. That suggests a group that contains U, s, and μ/ρ but not g. If the group is $U s^a (\mu/\rho)^\beta$, then by the same argument as before

$$\left[U s^a \left(\frac{\mu}{\rho} \right)^\beta \right] = [LT^{-1}][L^\alpha][L^2 T^{-1}]^\beta = [L^{1+\alpha+2\beta} T^{-1-\beta}] \qquad (4.5.11)$$

and therefore if the group is dimensionless β is -1 and α is $+1$, and the group is $\rho U s/\mu$. Fluid mechanics calls this group Reynolds number Re, after Osborne Reynolds (1842–1912), an engineer and professor at the University of Manchester, who clarified the concept of turbulence and measured the pressure drop in flow in pipes. Reynolds number reflects a ratio between forces related to inertia and forces related to viscosity, and appears in many fluid mechanics contexts.

Now taking viscosity into account

$$\frac{D}{\rho U^2 s^2} \quad \text{is a function of} \quad \frac{U}{\sqrt{(gs)}} \quad \text{and} \quad \frac{\rho U s}{\mu}$$

Returning to the idea of a physical model, we would like both $U/\sqrt{(gs)}$ and $\rho U s/\mu$ to be the same in the model as in the prototype, so that both

$$\frac{U_m}{\sqrt{g_m s_m}} = \frac{U_p}{\sqrt{g_p s_p}} \qquad (4.5.12)$$

and

$$\frac{\rho_m U_m s_m}{\mu_m} = \frac{\rho_p U_p s_p}{\mu_p}$$ (4.5.13)

where subscript m stands for model and p for prototype, as before. However that turns out to create difficulties. We saw earlier that we can readily meet the first condition (4.5.12), and that it leads to reasonably sized models that move at reasonable speeds in water. In principle, we could try to meet the second condition (4.5.13) as well as the first, by changing the liquid the model sails in, though a liquid other than water would create many practical difficulties. Dividing the left-hand side of (4.5.13) by the left-hand side of (4.5.12), and dividing the right-hand side of (4.5.13) by the right-hand side of (4.5.12), and then rearranging, meeting both conditions requires that

$$\frac{(\rho_m/\rho_p)}{(\mu_m/\mu_p)} = \left(\frac{s_p}{s_m}\right)^{3/2} \left(\frac{g_p}{g_m}\right)^{1/2}$$ (4.5.14)

A model on earth has to have g_p/g_m very close to 1. Continuing the example of the 1/50 scale model, the right-hand side of (4.5.14) is $50^{3/2}$, which is 354. The condition can only be met if we can find for the model a liquid that has a combination of very high density (to make ρ_m/ρ_p large) and very low viscosity (to make μ_m/μ_p small).

Unfortunately, there is no such liquid. The densest liquids are heavy metals such as mercury and molten uranium, but both create severe safety problems, and in any case their densities are only 10–20 times greater than water. The least viscous liquids are liquefied gases such as carbon dioxide, and low-boiling point hydrocarbons such as pentane, but they have rather low densities, are impractical for safety and other reasons, and their viscosities are smaller than water by a factor of less than 20. Equation (4.5.14) cannot be satisfied.

The way that this difficulty is resolved in practice is to keep water as the liquid, and to divide the total resistance into two components, one wave-making component related to inertia and gravity and the other related to viscosity. The second component can be more reliably calculated than the first. The model test is carried out so that (4.5.12) is satisfied (*Froude similarity*) but (4.5.13) is not. The total resistance is measured in the model, the calculated viscous component for the model is subtracted, and the remaining wave-making component is scaled up to the prototype in accordance with (4.5.6). The calculated viscous component for the prototype is then added in, to obtain the total resistance for the prototype. Rawson and Tupper[6] describe this in more detail.

4.6   Heat transfer from pipelines

Pipelines on the seabed carry oil between subsea wellheads and production platforms. The oil comes out of the subsurface reservoir at a high temperature, typically between 80 and 150 °C. The sea water is much colder, between 5 and 15 °C. Heat is conducted from the oil to the sea, and the oil temperature falls. If it falls too far, the oil becomes excessively viscous, and water and hydrocarbons in the oil may combine to form solid snow-like hydrates, which can block the pipeline. Many pipelines are insulated to avoid these problems. The insulation is expensive, and so it is desirable to minimize its volume.

Figure 4.4 shows a simple idealization of the problem. The pipe is steel. It is coated with a thin layer of anticorrosion coating, and then with a

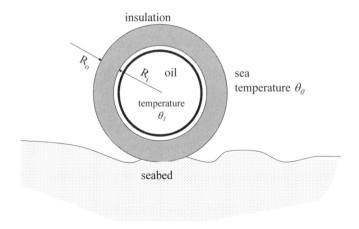

Fig. 4.4. Pipeline on seabed.

thicker layer of insulation. The outside of the insulation is in contact with the sea.

First, imagine that the only thermal resistance is provided by the insulation. This is a reasonable starting point: steel is a very good conductor of heat, and has a thermal conductivity of 40 W/m°C, whereas insulants have thermal conductivities smaller by a factor of about 200 to 5000. The anticorrosion coating has a thermal conductivity of about 0.2 W/m°C, but is only 2 or 3 mm thick. Assume as well that the thermal resistance between the sea and the outside of the insulation is negligible, and that the temperature distribution along the length of the pipeline has reached a steady state, so that the temperatures are not changing with time. We are interested in how the rate of heat transfer between the oil and the sea depends on the thermal conductivity and the dimensions.

A longer pipeline loses more heat that a shorter one, and the oil temperature falls as the oil flows from the upstream end to the downstream end. The rate of energy transfer q per unit length depends on the local difference $\theta_o - \theta_s$ between the oil temperature θ_o and the sea temperature θ_s, on the thermal conductivity k, and on the inside radius R_i and outside radius R_o of the insulation.

In the MLT system of fundamental units, energy has the dimension $[ML^2T^{-2}]$ of force $[MLT^{-2}]$ multiplied by distance $[L]$. If the conversion of mechanical energy to heat energy was significant in this problem, as it is in a brake that converts kinetic energy to heat, for instance, then we would need to give energy the dimension $[ML^2T^{-2}]$. But if energy conversion is not significant, we can choose to give heat energy its own fundamental dimension $[E]$. The dimension of q is then $[EL^{-1}T^{-1}]$, energy per unit length per unit time. Thermal conductivity k was discussed earlier in Sec. 2.2: it is energy per unit area per unit time per unit temperature gradient, and its dimension is

$$\left[\frac{\text{energy}}{(\text{area})(\text{time})(\text{temperature gradient})} \right] = \frac{[E]}{[L^2][T][\Theta/L]} = [EL^{-1}T^{-1}\Theta^{-1}]$$

(4.6.1)

where Θ is temperature, an independent dimension that cannot be obtained by combining E, L, and T. In an $ELT\Theta$ system of fundamental units, the dimensions of the different variables that enter the problem are therefore

heat flow rate per unit length q $\quad[EL^{-1}T^{-1}]$
thermal conductivity k $\quad[EL^{-1}T^{-1}\Theta^{-1}]$
temperature difference $\theta_o - \theta_s$ $\quad[\Theta]$
inside radius of insulation R_i $\quad[L]$
outside radius of insulation R_o $\quad[L]$

The relevant dimensionless groups can be determined either by inspection or by the systematic process described earlier. Proceeding by inspection, we see that [E] only comes into the first two variables, and that a dimensionless group that includes either of them must include both, and that they must appear as the ratio q/k. [Θ] only comes into the second and third variables k and $\theta_o - \theta_s$, in k as [Θ] and in $\theta_o - \theta_s$ as [Θ^{-1}], and so a dimensionless group that includes either of them must include both, and they must appear as the product $k(\theta_o - \theta_s)$. Combining these arguments, look for a dimensionless group of the form $q/k(\theta_o - \theta_s)R_o^\alpha$. It is dimensionless if α is 0, and so the group is $q/k(\theta_o - \theta_s)$.

The radii R_i and R_o have not yet been brought in, but their ratio R_i/R_o is dimensionless. Bringing this together, the dimensionless groups are $q/k(\theta_o - \theta_s)$ and R_i/R_o, and the first group must be a function of the second. A relationship between five variables has been reduced to a relationship between two dimensionless groups.

The argument can be taken a little further. If the layer of insulation is thin, so that R_i/R_o is close to 1, the rate of heat transfer must be inversely proportional to $R_o - R_i$, the thickness of the insulation. It follows that $q/k(\theta_o - \theta_s)$ is then linearly proportional to $1/(1 - R_i/R_o)$. A straightforward analysis shows that the constant of proportionality is simply 2π.

The problem can be generalized. Imagine now that there is some resistance to heat transfer across the outer boundary of the insulation. At that boundary the heat flow rate per unit area per unit time is the temperature difference between the surface of the insulation and the sea far away from the pipeline, multiplied by a surface heat transfer coefficient denoted h. The dimension of h is

$$\left[\frac{\text{energy}}{(\text{area})(\text{time})(\text{temperature difference})}\right] = \frac{[E]}{[L^2][T][\Theta]} = [EL^{-2}T^{-1}\Theta^{-1}]$$

(4.6.2)

If we want to retain the existing dimensionless groups but keep q out of a new group that includes h, the new group must include h/k, whose dimension is $[L^{-1}]$. It follows that hR_o/k is a suitable dimensionless group. There is nothing special about the choice of R_o rather than R_i: either would do equally well. Accordingly, the dimensionless groups are $q/k(\theta_o - \theta_s)$, R_i/R_o, and hR_o/k. The first group must be a function of the other two, and a relationship between six variables has been reduced to a relationship between three dimensionless groups. hR_o/k is a Nusselt number, written Nu, called after W. Nusselt (1882–1957), a pioneer of heat transfer theory.

4.7 Problems

4.1 Set up pendulums with different lengths and different masses, measure the period of each, and plot $t/\sqrt{s}$ against a/s on a graph. Observe what happens when a/s is small. If you take g as the 'standard' value 9.80665 m/s^2 (since on the Earth's surface variations between one place and another are very small), or (if you know it) the value of g where you do your experiments, what is the corresponding value of $t/\sqrt{s/g}$?

4.2 A spherical ball that falls into sand (or another granular material) embeds itself into the sand. Suppose the depth of embedment (measured from the original surface of the sand to the bottom of the ball) to depend only on

- the diameter of the ball
- the kinetic energy of the ball just before it strikes the sand
- the density of the sand
- the gravitational acceleration g

Carry out a dimensional analysis of this problem. Then find a source of dry sand (beach, children's sandpit, firebucket), and some suitable balls (ball bearing, marble, tennis ball, football), and do some experiments. The weight per unit volume of the sand can be measured by weighing a known volume (in a cook's measure or a bucket). Compare the experiments with your theory.

Use your results to make a very rough estimate of the kinetic energy and velocity of the meteorite that fell 15 million years ago and created the 25 km diameter Nördlinger Ries impact crater, 110 km north-west of Munich in Germany. Compare your energy estimate with the energy of a hydrogen bomb. Examine the assumptions you have had to make, and decide how you would proceed to make a better estimate.

Repeat the experiments with the same sand saturated with water. Are the differences explained by the difference in density?

4.3 The pendulum analysis in Secs. 4.1 and 4.3 considered the period t as a function of the length s, the gravitational acceleration g, the mass m, and the amplitude a. Repeat the analysis, now considering the length s as a function of t, g, m, and a. Are the conclusions different?

4.4 (Continuing from Problem 3.3) Water flows past a cylindrical pile that supports the deck of a loading pier. Vortex shedding is important in practice because as each vortex leaves the cylinder it gives it a small push. You can feel these forces if you move your finger through still water in a bath. Though the forces are small, they can induce large oscillations if the vortex shedding frequency is close to one of the natural frequencies with which the cylinder can oscillate, and if the damping is small.

Imagine that the vortex shedding frequency f can depend only on

- the pile diameter D
- the water velocity U
- the water density ρ

Apply dimensional analysis to show

(i) that the density must be irrelevant;
(ii) that f must be proportional to U/D.

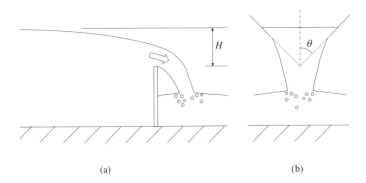

Fig. 4.5. *V-notch weir. (a) Section on centerline and (b) view looking upstream from below weir.*

Generalize the analysis to consider what happens if viscosity is important.

4.5   Figure 4.5 shows a V-notch weir, a simple and robust device for measuring water flow in an irrigation scheme. The volumetric flow rate q (volume/time) depends on the water density ρ, the gravitational acceleration g, the angle θ, and the vertical distance H between the bottom of the V and the surface of the water well upstream of the notch. Apply dimensional analysis to show that the flow rate must be proportional to $H^{5/2}$, for a given notch geometry defined by the angle θ.

4.6   The natural frequencies $N_1, N_2, N_3, \ldots$ of an organ pipe depend on the length of the pipe and the velocity of sound. Find the form of the relationship.

4.7   An air bubble oscillates radially in water with a natural frequency N. If the frequency depends only on the density of water, the local pressure, and the radius of the bubble, find the form of the relationship.

Sonar is used to locate fish, by sending an acoustic signal from a boat and detecting the reflections. The fish contain 'bubbles' in the form of swim bladders, whose volume they control so as to stay neutrally buoyant. The bladders reflect strongly at the natural frequency, and this can be used to estimate the size of the fish.[7]

If a bubble is very small, surface tension S (described in Problems 3.6 and 3.7 in Chap. 3) also has an effect on the natural frequency. Find the form that relationship takes.

4.8 The wavelength λ of small periodic surface waves on deep water depends only on the wave period t and the gravitational acceleration g. Apply dimensional analysis to find the form of the relationship.

If the waves are larger, the wavelength also depends on the wave height H. Apply dimensional analysis to find the form of the relationship.

If the waves are large and the wavelength is less than half the water depth, the wavelength depends on the water depth d as well as the wave height H. Generalize your analysis to cover this case.

4.9 (Continuing from Problem 3.1) Most pumps induce a flow rate that depends on the difference between the suction pressure (on the upstream side) and the discharge pressure (on the downstream side): if the pressure difference is larger, the flow is smaller. The volumetric flow rate, the volume of fluid pumped in unit time is q, the speed of rotation of the pump is N revolutions in unit time, the diameter of the pump rotor is D, the fluid density is ρ, and the pressure difference is p. Show that the relation between volumetric flow rate and discharge pressure can be expressed as a relationship between two dimensionless groups — one of them already derived in Problem 3.1 — and that this relationship will apply to any geometrically similar pumps.

4.10 There is a need to construct fixed offshore petroleum production structures in the Arctic Ocean, where there are important oil and gas fields off the north coast of Alaska, to the north of the Mackenzie Delta, near Novaya Zeml'ya and off the Yamal' peninsula. The major factor governing the design of these structures is the very large forces that can be exerted by drifting sea ice, which is 2 m thick, sometimes in pieces 100 km across containing ridges 30 m thick. The driving forces from wind and current are large, and the maximum design force on a structure is governed by the strength of the ice.

Suppose the ice force P to be determined by the contact area A between the ice and the structure, and that the relevant material parameter for the ice is the compressive strength Y, which has the dimensions of stress. Show by dimensional analysis that the maximum force per unit area P/A is independent of A.

Ice is a proverbially brittle material (which is why an ice cube out of the freezer cracks when you pour gin on it). That suggests an alternative model, that the relevant material parameter might not be compressive strength Y but might instead be fracture toughness K (see for example Ref. 8); K has the dimension $[FL^{-3/2}]$. Show that in that case P/A is not independent of A, and find how it depends on A.

Figure 4.6 plots indentation pressure P/A against contact area A for field and laboratory measurements over a wide range of scales. Each 'cloud' represents a large number of tests. The cloud marked 'lab' corresponds to model tests in a laboratory. The cloud marked 'field' represents tests on a larger scale, carried out in the Arctic with indenters pushed into large blocks of natural ice. The third cloud marked 'Hans Island' represents a large-scale experiment that Nature carries out for us. Hans Island lies in the Kennedy Channel between Greenland and Canada, and when the sea

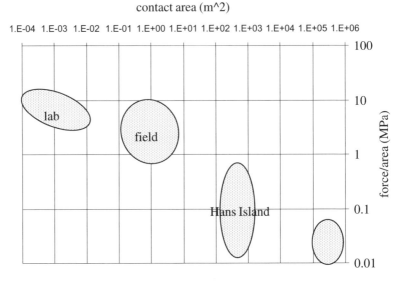

Fig. 4.6. Measurements of ice forces.

ice breaks up in July large ice floes, sometimes 3000 m across, drift down the Channel and occasionally strike the Island. The force between the island and the ice floe can be determined by installing accelerometers on the ice floe, measuring the deceleration as it strikes the island, estimating the mass of the floe, and applying Newton's law of motion. The fourth unlabelled cloud has a different status, and is not a direct measurement: it represents the ice force model that has to be put into a numerical model of the Arctic Ocean, in order to make the calculated motions match the observed motions. The original data are found in Sanderson[9] and Palmer and Sanderson.[10]

Which of the two models you have produced is more consistent with the data ?

Now suppose that P depends on A and on both K and Y. Use dimensional analysis to describe how to plot data in a nondimensional form which would allow you to investigate this dependence.

4.11   An atomic bomb exploded above the surface of the earth generates a rapidly expanding spherical shock wave. The explosion can be idealized as an instantaneous point release of a large quantity of energy. The radius r of the shock wave at time t after the explosion is a function of the energy release E, the initial density ρ of the atmosphere, and the ratio γ of specific heats at constant volume and constant pressure. Use dimensional analysis to determine how r depends on t.

The photographs below are from the Trinity explosion in New Mexico in July 1945: times are indicated, and the radius r can be scaled off the photos. Both sets of photos are of the same explosion. Plot log r against log t, and see if your result fits the data.

If the constant of proportionality between nondimensionalized radius and nondimensionalized energy is about 1 (which is no more than a guess, although as it happens a reasonable guess), estimate E. The density of air at 10 °C is 1.25 kg/m^3.

Think about the idealization of the explosion as a release of energy. Why is the mass of the bomb likely to be unimportant, except in so far as it determines the energy?

PLATE 1

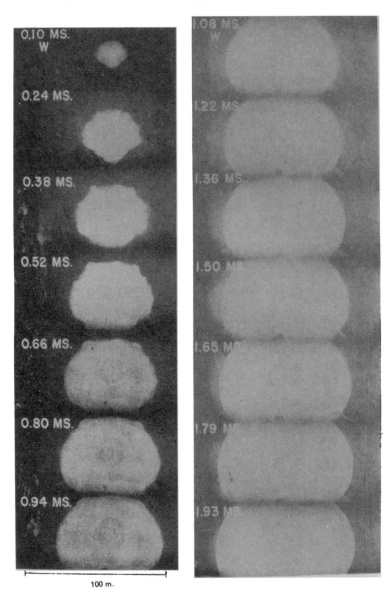

100 m.

Fig 6. Succession of photographs of the 'ball of fire' from $t = 0.10$ msec. to 1.93 msec.

Fig. 4.7. From Ref. 11, reproduced by permission of the Royal Society.

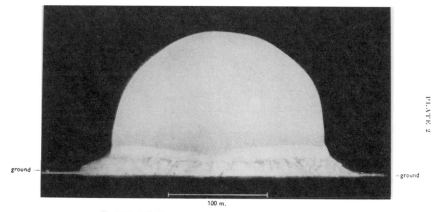

ground — | —ground

100 m.

Fig. 7 The ball of fire at $t = 15$ msec., showing the sharpness of its edge.

Fig. 4.8. From Ref. 11, reproduced by permission of the Royal Society.

The applied mathematician G.I. Taylor (1886–1975) used the photographs to calculate and publish the energy released by the bomb,[11] and thereby made himself unpopular in some quarters. His argument did not rely on dimensional analysis alone: why not? The incident appears in a novel by Tom McMahon, a professor of engineering at Harvard,[12] but Taylor took exception to the novel because of other aspects of the character modeled on him.[13]

4.12 Oil flows through a porous rock toward an oil well. The flow is slow enough to be laminar and governed by the oil viscosity. Other things being equal, the flow rate is proportional to the pressure gradient. One-dimensional flow in the x-direction is described by the equation

$$q = -\frac{k}{\mu}\frac{\partial p}{\partial x}$$

where q is the volume of oil per unit time that flows across unit area perpendicular to the x-direction; k a property of the porous rock, called the permeability, and depending on the size and shape of the pore, the extent of cementing between the particles, and the presence of water which is stuck to the pore sides and does not move; μ the viscosity of the oil; and $\frac{\partial p}{\partial x}$ is the pressure gradient in the x-direction.

Find the dimensions of k. If the rock is a sandstone made up of sand particles of more or less uniform size, how would you expect k to depend on the dimensions of the particles?

4.13 Heat transfer by convection happens when a fluid at one temperature flows across a solid object at a different temperature. It is called *forced* convection when the flow is driven externally.

Consider forced convection around a stationary object, at a temperature θ_o and characterized by a dimension s. Away from the object the free-stream velocity is U and the temperature θ_f. The fluid has density ρ, specific heat capacity c (the energy required to raise the temperature of unit mass by one temperature unit), thermal conductivity k, and viscosity μ. The heat transfer rate J is the amount of energy transferred between the body and the fluid in unit time. Take the fundamental dimensions as [E], [M], [L], [T], and [Θ], justifying taking [E] as an independent dimension by the argument used in Sec. 4.6, since conversion of kinetic energy to heat is insignificant in most convection situations.

 (i) Begin by determining the dimensions of the variables $s, (\theta_o - \theta_f), U, \rho, c, k, \mu$, and J. Only the temperature difference $\theta_o - \theta_f$ is taken as a variable, because it is the difference that creates the flow of heat: the absolute values are irrelevant.
 (ii) Determine how many independent dimensionless groups are there.
 (iii) Clearly there are many possible dimensionless groups. Look first for a simple dimensionless group that includes J and $\theta_o - \theta_f$, possibly using one of the groups in Sec. 4.6 as a starting point. Then look for another group that includes the free-stream velocity. Finally, find a dimensionless group that includes only the variables that describe the properties of the fluid, but not the body or the flow.
 (iv) Write down the relationship between the groups.
 (v) Consider what happens if kinetic energy can be converted into heat energy, so that energy has the dimension $[ML^2T^{-2}]$, instead of being treated as having a distinct dimension of its own.

4.14 (continuing from Problem 4.13) *Free* convection occurs when the heat transfer itself drives the flow, because changes of temperature in the fluid induce thermal expansion, and therefore local changes in the density and weight of the fluid. This sets up convection currents, and they carry away heated fluid and replace it with colder fluid from the surroundings. Convection is a highly efficient mechanism for heat transfer: a cup of coffee on the author's desk is cooling primarily by convection, whereas radiation and conduction to the desk play only a small part.

Denote by β the fractional change in volume of the fluid per unit change of temperature at constant pressure (analogous to the thermal expansion coefficient for a solid, but for volume rather than length). If the density is ρ, a unit temperature increase produces a change of density of $-\beta\rho$ and a change of unit weight of $-\beta\rho g$, where g is as usual the gravitational acceleration. That is the driving force per unit volume per unit temperature change, and so we need it as a variable. At a distance from the body the fluid is stationary, and so we no longer need U. The other seven variables are the same as in Problem 4.10.

(i) Observe that two of the three groups derived in Problem 4.8 can remain unchanged.
(ii) Derive a third group from the remaining variables, excluding J and either k or μ.

Rogers and Mayhew[14] discuss free and forced convection at much greater length. Beautiful photographs of streamlines in free convection are to be found in Van Dyke.[15]

4.15 A xylophone consists of a series of steel bars, which vibrate in bending and radiate sound when they are struck by a hammer. Each bar has an infinite number of natural frequencies, and each bar is supported at the nodal points for its lowest natural frequency. The nodal points for a natural frequency are the points that do not move when the bar is vibrating at that frequency. The higher frequencies have different nodal points, and so vibrations at those higher frequencies are more heavily damped and die away rapidly. The lowest natural frequency N for a bar depends on

— the length of the bar L
— the mass per unit length m
— the elastic flexural rigidity F

Find the form of the relationship between the variables.

Flexural rigidity is the ratio between the bending moment applied to a bar and the change of curvature the bending moment induces. Curvature is the reciprocal of radius of curvature. The dimensions of flexural rigidity are therefore

$$\left[\frac{\text{bending moment}}{\text{curvature}}\right] = \left[\frac{\text{force} \times \text{length}}{1/\text{radius of curvature}}\right] = \left[\frac{[\text{MLT}^{-2}][\text{L}]}{\text{L}^{-1}}\right]$$

$$= [\text{ML}^3\text{T}^{-2}]$$

References

1. Richardson, LF (1920). Convective cooling and the theory of dimensions. *Proceedings of the Physical Society of London*, 32, 405–409.

2. Vaschy, E (1892). Sur les lois de similitude en physique (On the laws of similarity in physics). *Annales Télégraphiques*, 19, 25–28.

3. Buckingham, E (1914). On physically similar systems: Illustrations of the use of dimensional equations. *Physical Review*, 4, 345–376.

4. Sterrett, SG (2005). *Wittgenstein Flies A Kite: A Story of Models of Wings and Models of the World*. New York: Pi Press.

5. Strang, G (1988). *Linear Algebra and Its Applications*. Florence, KY: Brooks/Cole.

6. Rawson, KJ and EC Tupper (1994). *Basic Ship Theory*. London, UK: Longman.

7. Dowling, AP and JE Ffowcs Williams (1983). *Sound and Sources of Sound*. Chichester: Ellis Horwood.

8. Ashby, MF and DRH Jones (1980). *Engineering Materials: An Introduction to their Properties and Applications*. New York: Pergamon Press.

9. Sanderson, TJO (1988). *Ice Mechanics: Risks to Offshore Structures*. London: Graham & Trotman.

10. Palmer, AC and TJO Sanderson (1991). Fractal crushing of ice and brittle solids. *Proceedings of the Royal Society, Series A*, 433, 469–477.

11. Taylor, GI (1950). The formation of a blast wave by a very intense explosion. II. The atomic explosion of 1945. *Proceedings of the Royal Society, series A*, 101, 175–186.

12. McMahon, T (1970). *A Random State*. New York: Macmillan. In the USA published as *Principles of American Nuclear Chemistry: A Novel*. Macmillan (1970) and University of Chicago Press (2003).

13. Batchelor, G (1996). *The Life and Legacy of G.I. Taylor*. Cambridge: Cambridge University Press.

14. Rogers, GFC and YR Mayhew (1980). *Engineering Thermodynamics, Work and Heat Transfer*. London: Longman.

15. Van Dyke, M (1982). *An Album of Fluid Motion*. Stanford: Parabolic Press.

Chapter 5

Similarity and Intelligent Experimentation

5.1 Similarity

Rayleigh made similarity arguments the basis of his famous statement quoted in Chap. 1:

> *"It happens not infrequently that results in the form of 'laws' are put forward as novelties on the basis of elaborate experiments, which might have been predicted a priori after a few minutes consideration."*

Dimensional analysis can be used to assemble the results of experiments in a concise and accessible form, so that we can arrive at broadly applicable general results from a small number of tests, often at a model scale.

For example, engineers frequently want to transmit fluids along pipes, and need to know the pressure difference required to generate a required flow rate. The pipes have very different diameters, from several meters (penstocks in hydroelectric plants, magma channels in volcanoes) down to small fractions of a millimeter (microfluidic devices, blood vessels in the brain), and the fluids are very different (air, water, oil, liquid sodium, sewage). It would be extraordinarily awkward if every design calculation for a fluid in a pipe had to be based on a previous experiment on a pipe with the same fluid and the same diameter. Dimensional analysis tells us that we can carry out experiments with one diameter and one fluid at different velocities, and that if we analyze the results in the right way we can derive a universal curve that we can apply to any diameter and any fluid. This example is discussed in Sec. 5.2. Section 5.3 describes a different example.

5.2 Pressure drops in pipes

A pressure difference is required to push a fluid through a pipe. If we want the fluid to flow at a required rate, we need to be able to calculate the pressure difference between the ends of the pipe, so that we can be sure that enough pressure is available, and to size pumps if they are needed.

First, consider the steady flow of a Newtonian fluid through a horizontal pipe, and suppose the pipe to be circular and the inside surface of the pipe to be perfectly smooth. 'Newtonian' means that the shear strain rate in the fluid is proportional to the shear stress: this is discussed further in Sec. 4.5. Most but not all of the fluids engineers and scientists work with are Newtonian.

Distance along the pipe is measured by x. The difference between the pressure p at the upstream and downstream ends generates a pressure gradient dp/dx, the rate of change of pressure with distance. Pressure is force per unit area, and so its dimensions are [force/area] which is $[MLT^{-2}/L^2] = [ML^{-1}T^{-2}]$. The dimensions of pressure gradient are therefore $[ML^{-2}T^{-2}]$. The pressure gradient at a particular point on the pipeline depends on the local conditions at that point, and is not necessarily the same as the pressure gradient at other points.

Suppose that the pressure gradient is a function of the local internal diameter D, the local mean velocity U, the local fluid density ρ, and the local viscosity μ. The mean velocity is the velocity averaged over the internal cross-section, and is equal to the volume flowing past a fixed point in unit time, divided by the cross-sectional area. Density comes in because at high speeds the streamlines are not parallel to the pipe axis, but instead the flow is turbulent, and the flow includes many eddies of different sizes, in which the fluid is accelerated and decelerated. Viscosity is defined and discussed in Sec. 4.5; we expect it to be significant because honey or oil will require a larger pressure gradient than water or mercury, for the same flow rate and pipe diameter.

The variables and their dimensions are

pressure gradient dp/dx $[ML^{-2}T^{-2}]$
diameter D $[L]$
velocity U $[LT^{-1}]$
density ρ $[ML^{-3}]$
viscosity μ $[ML^{-1}T^{-1}]$

A dimensionless group based on the first four parameters is $(dp/dx)D^{\alpha}U^{\beta}\rho^{\gamma}$, whose dimensions are

$$[ML^{-2}T^{-2}][L]^{\alpha}[LT^{-1}]^{\beta}[ML^{-3}]^{\gamma} = [M^{1+\gamma}L^{-2+\alpha+\beta-3\gamma}T^{-2-\beta}]$$

and this group is dimensionless when

$$\begin{aligned}
[M] \quad & 0 = 1 + \gamma \\
[L] \quad & 0 = -2 + \alpha + \beta - 3\gamma \\
[T] \quad & 0 = -2 - \beta
\end{aligned}$$ (5.2.1)

and so from the first and third of the three equations

$$\gamma = -1$$ (5.2.2)

$$\beta = -2$$ (5.2.3)

and substituting into the second

$$\alpha = 1$$ (5.2.4)

and so the dimensionless group is $(dp/dx)D/\rho U^2$.

The effect of viscosity can be taken into account by a second dimensionless group that includes the viscosity μ. There is more than one way of making such a group, but if we want to include μ but not include dp/dx we can apply the same arguments used in Sec. 4.5. The group is then Reynolds number Re, $\rho UD/\mu$.

A general relation between five variables has been reduced to a relationship between two dimensionless groups, so that

$$\frac{(dp/dx)D}{\rho U^2} \quad \text{is a function of } \rho UD/\mu$$

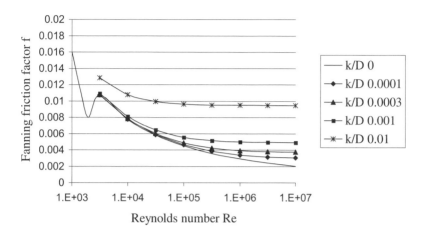

Fig. 5.1. *Relation between friction factor and Reynolds number.*

Everything we need for Newtonian fluids in smooth pipes can be summarized in a single graph that plots $(dp/dx)D/\rho U^2$ against $\rho UD/\mu$. This is done in Fig. 5.1, but there the ordinate is $(1/2)\,(dp/dx)D/\rho U^2$, called Fanning friction factor f. The discussion relates to a perfectly smooth pipe, and that corresponds to the curve labeled $k/D\,0$.

Notes:

1. That the scale for Re is logarithmic;
2. That on the right of the graph, where Re is large, friction factor f is independent of Re;
3. That on the left, where Re is small, f is proportional to $1/\text{Re}$, which makes the pressure gradient independent of ρ; as we shall see later;
4. That there is an upward jump in the graph when Re reaches about 2500: it would be grossly incorrect to carry out experiments at Re less than 2500 and imagine that the graph could be extrapolated;
5. That because of the jump, the relationship cannot be expressed as a power series.

The third of these results can be secured independently, again by dimensional analysis. It seems reasonable to suppose that if the flow is slow enough, the density of the fluid becomes irrelevant, because inertia has a negligible effect. The relevant variables are now

$$
\begin{array}{ll}
\text{pressure gradient } dp/dx & [ML^{-2}T^{-2}] \\
\text{diameter } D & [L] \\
\text{velocity } U & [LT^{-1}] \\
\text{viscosity } \mu & [ML^{-1}T^{-1}]
\end{array}
$$

A dimensionless group based on the this alternative choice of four parameters is $(dp/dx)D^{\alpha}U^{\beta}\mu^{\gamma}$, whose dimensions are

$$[ML^{-2}T^{-2}][L]^{\alpha}[LT^{-1}]^{\beta}[ML^{-1}T^{-1}]^{\gamma} = [M^{1+\gamma}L^{-2+\alpha+\beta-\gamma}T^{-2-\beta-\gamma}]$$

and so

$$
\begin{array}{lll}
[M] & 0 = 1 + \gamma & \\
[L] & 0 = -2 + \alpha + \beta - \gamma & \quad (5.2.5) \\
[T] & 0 = -2 - \beta - \gamma &
\end{array}
$$

whose solution is

$$\alpha = 2$$
$$\beta = \gamma = -1$$

and so the dimensionless group is $(dp/dx)D^2/\mu U$. If no other parameters enter the problem, $(dp/dx)D^2/\mu U$ must be a constant, denoted C, and then the friction factor f is

$$f = \frac{(1/2)\frac{dp}{dx}D}{\rho U^2} = \frac{1}{2}\frac{\frac{dp}{dx}D^2}{\mu U}\frac{\mu}{\rho U D} = \frac{1}{2}\frac{C}{\text{Re}} \qquad (5.2.6)$$

and in that regime f is inversely proportional to Re. That is what is observed in Fig. 5.1 at values of Re up to 2000. When Re is more than 2000, the flow becomes turbulent, the density becomes important, and (5.2.6) no longer holds.

Another extreme case occurs if the pressure gradient is taken to be independent of the viscosity rather than the density. There are then four variables.

pressure gradient dp/dx $[ML^{-2}T^{-2}]$
diameter D $[L]$
velocity U $[LT^{-1}]$
density ρ $[ML^{-3}]$

rather than five, and there is one independent dimensionless group $(dp/dx)D/\rho U^2$ which we derived earlier. It is not obvious that this case applies to highly turbulent flow at large velocities, because turbulent flow is always going to induce large velocity gradients and significant shear stresses, but Fig. 5.1 shows that at very high values of Re friction factor is indeed independent of Re.

These conclusions can again be generalized.

Until now it has been assumed that the inside surface of the pipe wall has no effect on the pressure gradient, but we would expect that a smooth surface (such as bare steel or a polymeric internal coating) would induce less resistance to flow than a rough surface (such as rusty steel, concrete, wax deposits, or fouling in water pipelines by hard or soft organisms such as mussels or sponges). A complete characterization of the roughness of a surface is a complex task that requires more than one geometric parameter. The simplest approach is to represent the roughness by a single length, called the roughness height k, with dimension [L], which is larger for a rough surface than for a smooth surface.

We now have six parameters instead of five, and there are still three fundamental dimensions, and so there are three dimensionless groups. There is no reason to abandon the two groups we already have. The obvious new dimensionless group is the ratio k/D, called the dimensionless roughness. The earlier result can be generalized to

$$\frac{(dp/dx)\,D}{\rho U^2} \quad \text{is a function of} \quad \frac{\rho UD}{\mu} \text{ and } \frac{k}{D}$$

Figure 5.1 includes a festoon of curves, each for a different value of k/D. If Re is less than 2000, the curves coincide, because there k has no effect (unless it is comparable with the diameter D). Only one chart is required, and the results are still tractable. The engineer who wants to calculate the

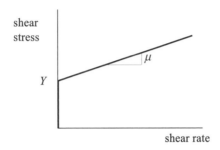

Fig. 5.2. *Relation between shear rate and shear stress for a Bingham fluid.*

pressure gradient that corresponds to a given flow rate determines Re (from the mean velocity, the pipe diameter, and the fluid properties ρ and μ) and k/D (from an assessment of the pipe surface), and can then determine $(dp/dx)D/\rho U^2$ from the chart, and multiply $(dp/dx)D/\rho U^2$ by $\rho U^2/D$ to find dp/dx.

An alternative generalization is to consider liquids that are not Newtonian, and that have a more complex nonlinear relationship between shear stress and shear rate. This is another vast subject, called rheology, but one simple idealization is illustrated in Fig. 5.2. In this idealization the shear rate is 0 if the shear stress is less than a yield stress Y. As long as the shear stress is less than Y, the liquid behaves as a solid and does not deform. If the shear stress is greater than Y, the liquid deforms at a rate proportional to the difference between the shear stress and Y, and the constant of proportionality between the shear stress and that difference is the viscosity μ.

This is called the Bingham idealization, and is useful for 'thick' liquids like heavy oil, bitumen, molten chocolate, and tomato ketchup. Now the sixth parameter is Y, a stress with dimension [force/area] which is $[ML^{-1}T^{-2}]$, and there are three dimensionless groups. Again there are choices. We might think that it would be useful to have a group that reflects the relative importance of yield stress effects that depend on Y and viscosity effects that depend on μ. the dimension of μ was derived earlier, and is $[ML^{-1}T^{-1}]$. The dimension of the ratio Y/μ is therefore $[T^{-1}]$. None of the parameters have dimension [T] on its own, but the ratio D/U does have dimension [T]. It follows that $YD/\mu U$ is dimensionless, and therefore that

$$\frac{(dp/dx)D}{\rho U^2} \text{ is a function of } \frac{\rho U D}{\mu} \text{ and } \frac{YD}{\mu U}$$

$\rho U D/\mu$ is Reynolds number again. $\rho U D/\mu$ and $YD/\mu U$ are both dimensionless and both include μ, but we can construct another group that does not include μ if we divide $\rho U D/\mu$ by $YD/\mu U$ to get $\rho U^2/Y$. That is Cauchy number, a group that often arises in impact engineering, a ratio between forces related to strength and forces related to inertia, in fluid of density ρ moving with velocity U.

There is an interesting special case. If the liquid can carry a nonzero shear stress, there can be a pressure gradient but no flow. The flow can only start if a critical pressure gradient is reached. Suppose, for example, that an insulated pipeline carries heavy oil at a high temperature, but that a pump breakdown stops the flow. The oil cools down and approaches the temperature of the surrounding environment. The cold heavy oil behaves as a Bingham fluid, and a finite pressure gradient is required to get the oil flowing again.

That critical pressure gradient to restart flow can depend only on the yield stress Y and the pipe diameter D. U is 0, and ρ and μ cannot be involved because the oil is not yet moving. The variables are

critical pressure gradient dp/dx	$[ML^{-2}T^{-1}]$
yield stress Y	$[ML^{-1}T^{-1}]$
diameter D	$[L]$

There are three fundamental dimensions, and three variables, and so we expect there to be no dimensionless group. However, this is an example of the case described in Sec. 4.2 that occurs when the dimension matrix $\mathbf{D}$ is singular. The matrix is

$$\mathbf{D} = \begin{bmatrix} 1 & 1 & 0 \\ -2 & -1 & 1 \\ -1 & -1 & 0 \end{bmatrix} \tag{5.2.7}$$

and can be seen to be singular by evaluating its determinant and finding it to be 0, or more rapidly by adding the third row to the first row, which generates a row of zeroes. The dimension matrix has rank r equal to 2 rather than 3, and so the number of independent dimensionless groups is $n - r = 3 - 2 = 1$. That happens because M and T appear only together, in the combination MT^{-1}, so that we could call MT^{-1} a new fundamental dimension X for the purpose of this problem, and then the dimensions would be

critical pressure gradient $(dp/dx)_{cr}$ $[XL^{-2}]$
yield stress Y $[XL^{-1}]$
diameter D $[L]$

and the dimension matrix would be

$$\mathbf{D} = \begin{bmatrix} 1 & 1 & 0 \\ -2 & -1 & 1 \end{bmatrix} \tag{5.2.8}$$

The dimensionless group is $D(dp/dx)_{cr}/Y$. There is nothing else for it to be a function of, and so it must have a fixed value. Simple theory (see, for example, Palmer and King[1]) shows that value to be 4.

5.3 Plows in clay

Plows are thought to have been invented in China some four thousand years ago. An agricultural plow breaks up and turns over the surface soil, so that water, air, and plant roots can move through it more easily. A plow can also be used to cut trenches. The plows that will be discussed here cut trenches 1 or 2 m deep in the seabed. The purpose of the trench is to protect a seabed pipeline against disturbance from currents, cables, and fishing gear.[2]

Figure 5.3 shows one plow of this kind, in a trial on shore. The plow is pulled by a tug or a moored barge. The front end rests on skids, and a beam connects the skids to a share that cuts and lifts the soil, and to moldboards that push the soil sideways so that it does not fall back into the trench. The purpose of this configuration is to secure automatic stable control of the trench depth, so that the plow neither digs in until it becomes an anchor nor rises up and scrapes along the surface. A horizontal heel plate attached under the share secures this control.

Fig. 5.3. Trenching plow.

The geometry of a plow is clearly quite complicated, and it deforms the soil in a complex way that geotechnics finds difficult to analyze. It is beyond the grasp of finite-element analysis, though that situation is beginning to change. It would be awkward and expensive to have to construct a series of full-scale plows in order to optimize the geometry. It is much cheaper to build and test model plows, but it is necessary to be sure that the model plow will behave in the same way as a geometrically similar full-scale plow, so that the results of the tests can be scaled up to make reliable predictions.

The weight of the soil is important, because the plow has to lift the soil out of the trench. The strength of the soil is important, because the plow has to cut and deform it. In clay soils, the relevant strength parameter is the undrained shear strength, denoted c: undrained means that there is no time for water to redistribute itself within the soil as the plow cuts through it. Geometrically similar plows can be characterized by one length, since all the other lengths scale in proportion, and it is convenient to take the trench depth, which is the vertical distance between the bottom of the skids and the bottom of the heel plate. The weight of the plow may be important: if the plow is heavy, it will sink into soft soil, and the weight may also be significant in the balance between soil forces and pull forces that controls the equilibrium of the plow. We are also interested in, the force required to pull the plow forward, the draft (a term taken from agricultural engineering), so that we can size the pulling equipment and design the plow so that it will be strong enough to resist the forces that act on it.

If the plow is moving slowly, inertia forces are not significant. The fundamental dimensions can then be taken simply as force F and length L, and the dimensions of the variables are

$$
\begin{array}{ll}
\text{soil weight per unit volume } \gamma & [FL^{-3}] \\
\text{soil undrained shear strength } c & [FL^{-2}] \\
\text{trench depth } H & [L] \\
\text{weight } W & [F] \\
\text{draft } D & [F]
\end{array}
$$

There are two dimensions and five variables, and therefore three independent dimensionless groups. If we had taken the fundamental dimensions as M, L, and T, we would have had one more dimension, but the dimension matrix would have had rank 2 rather than 3, and so there would still be three independent dimensionless groups.

The groups can be defined in different ways. A useful group is $c/\gamma H$, a group previously developed in Problem 3.4, recalling that weight per unit volume, denoted ρg there, is the same as γ here. The solution and discussion of that problem show that $c/\gamma H$ can be thought of as a ratio between two kinds of force, the force required to overcome the shear strength of the soil and the force generated gravity.

If we keep $c/\gamma H$ as the only group that includes soil weight γ, the other two independent groups are W and D divided by cH^2, and the three groups are

$$
\frac{c}{\gamma H}, \quad \frac{W}{cH^2}, \quad \text{and} \quad \frac{D}{cH^2}
$$

which indicates that a model test should be carried out in such a way that

1. The model plow is geometrically similar to the prototype plow,
2. $c/\gamma H$ is the same in the model as in the prototype, so that the ratio between gravity and strength is correct, and

3. W/cH^2 is the same in the model as in the prototype, so that the ratio between plow weight and strength is correct.

If we do that, D/cH^2 will be the same in the model as in the prototype, and we can use measurements of D in a test to predict D in a prototype.

Consider the practicability of making $c/\gamma H$ the same in the model as in the prototype. Denoting as before model by subscript m and prototype by subscript p, the condition is

$$\frac{c_m}{\gamma_m H_m} = \frac{c_p}{\gamma_p H_p} \tag{5.3.1}$$

and so

$$\frac{c_m/\gamma_m}{c_p/\gamma_p} = \frac{H_m}{H_p} \tag{5.3.2}$$

Accordingly, in a one-tenth scale model ($H_m/H_p = 1/10$) the ratio c/γ must have one-tenth the value in the model that it does in the prototype. The model soil has to be much weaker than the prototype soil, and this can be achieved by mixing water into the soil, which creates a big reduction in c but a much smaller reduction in γ.

The third condition is more difficult. The results of the test program will be used in the design, and so when we do the tests we do not know how heavy the prototype is going to be. The only solution is to make an intelligent guess at the prototype weight, scale the model weight accordingly, use the results to prepare a better estimate of the prototype weight, and if necessary repeat the test. It will clearly be desirable to make the weight adjustable.

This approach was first used to help design a plow that would cut a 1.2 m deep underwater trench in stiff clay in the Statfjord oilfield in the Norwegian sector of the North Sea.[2] Trench stability was not an issue, but it was very important to predict the force that would be needed to pull the plow, because it was intended to pull the plow with a tug (and if possible at the same time to pull a pipeline behind it, though in the end that was not done). Several one-tenth scale models were tested first, and different share and moldboard shapes were tried out to reduce the pull force as far as possible, and to move the plowed-out soil smoothly to the sides of the trench. The results of those tests were applied to a one-quarter scale model, and the design, testing and deployment of the full-scale plow followed.

Palmer *et al.*[3] describe in detail a second test program that applied these ideas to design a plow that would cut a 1.5 m deep trench in soft mud off the coast of Melville Island in Arctic Canada. The mud was so soft that the trench was not far from the limit at which it would collapse under its own weight, and the mud also had to support the plow's weight and the reaction from the skids and the heel. The plow weight had to be kept down so that the plow could be transported by air.

At the site where the full-scale plow was to be operated, c is 3 kPa (although there is a variation with depth below bottom, as well as signif-icant uncertainty about the precise value[4]) and γ is 7 kN/m^3 (taking the relevant weight per unit volume as the submerged weight, since the plow is to operate under water), and so c/γ is 0.43 m. A one-tenth scale model ought therefore to operate in soil with c/γ 0.043 m.

The first model tests were carried out above water on a tidal mud-flat. The mud has a strength and density roughly the same as those at the prototype site, and is therefore too strong to be correctly scaled, but the flat made it possible to carry out a large number of tests, and to observe the behavior of the plow. Those tests suggested some modifica-tions to the design, among them changing the skid supports, dividing each mouldboard into two sections hinged together and separately supported on skids, and pivoting the heel to the share. Figure 5.4 shows the modified model plow.

A modified design was then tested in correctly scaled conditions under water in a laboratory soil tank. It was not practicable to bring sev-eral tonnes of clay from the Arctic to the laboratory, and so a model clay was made by mixing water into another clay from a local fireclay manu-facturer, to bring c down to between 0.18 and 0.75 kPa. Estimates indicated that the prototype plow would weigh about 180 kN in water, and than the corresponding scaled weight ought to be 120 N. Because the model was heavier that that, part of its weight was taken by a counter-balance system. It was found that when the soil was very weak and the plow was heavy the rear end of the plow would sink into the mud, and the

Fig. 5.4. Model plough with floating mould board.

trench depth was reduced. This was controlled by adding buoyancy to the rear end.

The full-scale plow was tested on the bottom of a channel in the Fraser River delta in Western Canada. The site was not ideal, but there were many practical constraints. The plow was then taken to pieces, flown to Melville Island, reassembled, and pulled by a cable from a fixed winch on the ice to excavate the trench. The trench was 300 m long and 1.5 m deep. The total pull force was 120 kN, but nearly half of that was the pull required for the 1100 m pull cable, and the estimated draft for the plow was 66 kN.

Figure 5.5 plots D/cH^2 against $c/\gamma H$ for both test series, the field trials, and the final operation. It can be seen that the model-scale and full-scale data are consistent, and that the model tests were a good predictor of the force required to pull the full-scale plow. There is a large amount of scatter, and a few tests give very high values of D/cH^2: the reason is thought to be uncertainty about the values of both c and H. Toward the left of the diagram D/cH^2 is high: that happens because when $c/\gamma H$ is small the trench is only marginally stable: the theoretical limiting value is 0.17 for a

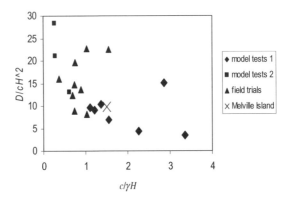

Fig. 5.5. Measured relation between dimensionless shear strength and dimensionless draft.

V-shaped trench with 45° sides, and that value does not allow for loading by the plow itself or by the spoil left along the sides of the trench.

A limitation of the analysis is that it does not account for the variation of shear strength with depth below the sea bottom, which tests on the field test site and at Melville Island showed to be significant.[3,4] A more refined analysis might include two shear strength parameters, the shear strength at the bottom and the shear strength at a distance H below the bottom.

It was lucky that the first two applications of the plow were in clay, where the tests showed that the plow draft is almost independent of plow speed, at any reasonable speed. In sand, on the other hand, there is a very substantial speed effect that scales differently, starts at rather small speeds, and is strongly dependent on trench depth.[5]

5.4 Problems

5.1 An undersea pipeline consists of a steel tube, covered by a thin layer of anticorrosion coating and often by a layer of concrete. The pipeline is initially laid on the seabed filled with air at atmospheric pressure, so that it is loaded by the external hydrostatic pressure of the sea: the process is described in Sec. 6.3. If the pipeline is bent severely in deep water, it can

Fig. 5.6. Buckled pipeline.

happen that it buckles under the combined action of bending and pressure, and that once it has been initiated the buckle runs along the pipeline, squashing it into the dumbbell cross-section seen in Fig. 5.6. That raises the frightening possibility that buckle propagation from a mishap during construction might lead to the loss of a great length of pipeline.

   The minimum external pressure at which a buckle can run along the pipeline is called the propagation pressure. It depends on the diameter D and wall thickness t of the pipe, and the yield stress Y of the steel. The effect of the coatings is negligible.

Propagation pressure can be measured by making a dent in the pipe to serve as an initiator, placing the pipe in a pressure vessel, pumping water into the annular space between the pipe and the pressure vessel, and measuring the change in internal volume of the pipe. Plotting pressure against change of volume, there is a peak as the buckle starts, and then the pressure settles down to a steady value at which the buckle moves along the pipe. Some of the available data are:

D	mm	827.0	827.4	827.2	827.1	777.9	332.1	331.7	331.5	331.7	332.4	331.7	331.8
t	mm	19.0	19.4	19.2	19.1	15.9	8.1	7.7	7.5	7.7	8.4	7.7	7.8
Y	MPa	467	490	512	484	426	365	359	331	331	336	313	357
p	MPa	1.22	1.22	1.21	1.26	0.76	1.17	1.00	1.07	0.93	1.20	0.76	0.95

D	mm	12.7	12.7	12.7	12.7	25.4	25.4	28.7	28.6	23.6	24.3	25.4
t	mm	0.25	0.20	0.20	0.46	0.56	0.71	0.98	1.22	0.31	0.71	1.22
Y	MPa	662	650	650	471	379	324	480	490	252	183	349
p	MPa	1.09	0.70	0.75	4.30	0.95	1.40	2.30	4.90	0.18	0.78	3.90

Some of the data are from Martin[6] and Kamalarasa and Calladine.[7]

Use dimensional analysis to plot these data on a single graph. Use the graph to predict the propagation pressure for a pipe with mean diameter 793.75 mm, wall thickness 19.05 mm, and yield stress 448.2 MPa, and calculate the corresponding depth of seawater.

There is some scatter, and it might be that the propagation pressure also depends on the elastic modulus (Young's modulus) of the steel as well as the yield stress Y. Decide how you would generalize the analysis to take account of this.

5.2 Small bubbles in *Coca-Cola* or champagne remain nearly spherical as they rise toward the surface. Larger bubbles take up a different shape, an irregular cap like a slice cut from a sphere. Taylor[8] measured the rate of rise of bubbles in nitrobenzene (density 1200 kg/m^3, viscosity 1.8×10^{-3} Pa s), and found the following data:

volume V	cm^3	1.48	3.50	4.06	4.31	6.40	7.30	8.80	9.18	18.4	21.3	33.8
rise velocity U	cm/s	29.2	29.6	28.9	28.0	35.5	34.2	37.2	33.0	37.3	38.1	42.1
maximum breadth	cm	2.86	3.14	3.48	3.16	4.98	4.40	5.26	4.53	5.10	5.85	6.19

Carry out a dimensional analysis, plot the data in whatever form seems the most instructive, and comment.

References

1. Palmer, AC and RA King (2004). *Subsea Pipeline Engineering*. Tulsa, OK: Pennwell.

2. Brown, RJ and AC Palmer (1985). Submarine pipeline trenching by plows. *Proceedings of the Seventeenth Annual Offshore Technology Conference*, Houston, Vol. 2, pp. 283–291.

3. Palmer, AC, JP Kenny, MR Perera and AR Reece (1979). Design operation of an underwater pipeline trenching plough. *Geotechnique*, 29, 305–322.

4. Palmer, AC (1979). Application of offshore site investigation data to the design and construction of submarine pipelines. Proceedings of the Society of Underwater Technology Conference on Offshore Site Investigation, London, pp. 257–265.

5. Palmer, AC (1999). Speed effects in cutting and plowing. *Geotechnique*, 49 (3), 285–294.

6. Martin, JH (1974). Propagating buckles in thin-walled tubes. Unpublished project report. Cambridge University Engineering Department.

7. Kamalarasa, S and CR Calladine (1988). Buckle propagation in submarine pipelines. *International Journal of Mechanical Sciences*, 30, 217–228.

8. Taylor, GI (1950). The mechanics of large bubbles rising through extended liquids and through liquids in tubes. *Proceedings of the Royal Society, Series A*, 200, 375–390.

Chapter 6

Equations in Nondimensional Form

6.1 Motivation

In the examples discussed in Chaps. 2, 3, and 4, we had an idea of what variables might be important to a problem, but we did not carry out any further analysis before embarking on dimensional analysis. We did not need to know what the governing equations might be, we made no attempt to derive them, and we were still able to arrive at useful results. Sometimes, though, we have the equations, and then we can take a parallel but different path by rewriting the governing equations in terms of dimensionless quantities. That often leads to useful general results

6.2 Mass–spring system

First think about the simple system shown in Fig. 6.1. A mass m moves horizontally on massless wheels. It is constrained by a linear spring. The force induced in the spring is its extension multiplied by a stiffness k. The horizontal displacement of the mass is X, measured from the equilibrium position at which the tension in the spring is 0. When the displacement is X to the right, the spring tension is kX, and that force pulls the mass to the left. The acceleration of the mass is d^2X/dT^2, where T is time. The equation of motion for the mass is

$$m\frac{d^2X}{dT^2} = -kX \qquad (6.2.1)$$

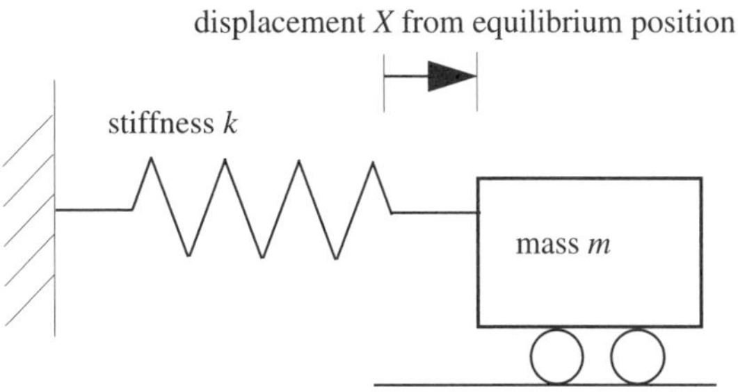

Fig. 6.1. Mass–spring system.

Instead of solving the equation immediately, we change X and T to dimensionless variables x and t. Let

$$t = \frac{T}{\sqrt{m/k}} \qquad (6.2.2)$$

and

$$x = \frac{X}{a} \qquad (6.2.3)$$

where a is an arbitrary length, so that

$$T = t\sqrt{m/k} \qquad (6.2.4)$$

and

$$X = ax \qquad (6.2.5)$$

Here and from now on we adopt the useful convention that dimensionless variables like t and x are represented by lower-case symbols, and the corresponding dimensioned variables T and X are represented by capitals. Now use (6.2.4) and (6.2.5) and substitute for X and T in (6.2.1), which becomes

$$m\frac{d^2(ax)}{d(t\sqrt{m/k})^2} = -k(ax) \qquad (6.2.6)$$

which is

$$ma\frac{1}{m/k}\frac{d^2x}{dt^2} = -kax \qquad (6.2.7)$$

which simplifies (by dividing by ka and canceling the m's on the left) into

$$\frac{d^2x}{dt^2} = -x \qquad (6.2.8)$$

The quantities $m, k,$ and a have disappeared and we now have a single equation in two dimensionless variables. Even if we cannot solve the equation, we may be able to make useful statements about it, and they will apply to any system of the form sketched in Fig. 6.1. If we can solve the equation, we have a solution that applies universally, whatever the values of $m, k,$ and a.

The general solution to the second-order differential equation (6.2.8) is elementary, and is

$$x = b \sin t + c \cos t \qquad (6.2.9)$$

where b and c are arbitrary constants. The solution is periodic, with period 2π in the dimensionless time t, because $\sin(t + 2n\pi) = \sin t$ and $\cos(t + 2n\pi) = \cos t$ for any t and any integer n. Going back to the original values by substituting from (6.2.2) and (6.2.3), the general solution is

$$X = B \sin\left(T\sqrt{\frac{k}{m}}\right) + C \cos\left(T\sqrt{\frac{k}{m}}\right) \qquad (6.2.10)$$

and the period is $2\pi\sqrt{m/k}$).

Next think of a generalization sketched in Fig. 6.2. In Fig. 6.1 the only force that acts on the mass was the spring tension. In Fig. 6.2 an external periodic force $P \sin \omega T$ has been added. The equation of motion for the mass becomes

$$m\frac{d^2X}{dT^2} = -kX + P \sin \omega T \qquad (6.2.11)$$

Again change X and T to dimensionless variables x and t. Let

$$t = \omega T \qquad (6.2.12)$$

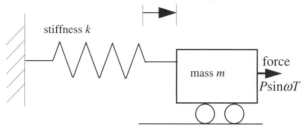

displacement X from equilibrium position

stiffness k

mass m

force $P\sin\omega T$

Fig. 6.2. *Mass–spring system with external force.*

and

$$x = \frac{X}{P/k} \tag{6.2.13}$$

so that

$$T = t/\omega \tag{6.2.14}$$

and

$$X = Px/k \tag{6.2.15}$$

Substituting into (6.2.11) it becomes

$$m\frac{d^2\left(\frac{Px}{k}\right)}{d\left(\frac{t}{\omega}\right)^2} = -k\left(\frac{Px}{k}\right) + P\sin t \tag{6.2.16}$$

which simplifies to

$$\left(\frac{\omega}{\omega_n}\right)^2 \frac{d^2x}{dt^2} = -x + \sin t \tag{6.2.17}$$

where in addition

$$\omega_n = \sqrt{k/m} \tag{6.2.18}$$

Equation (6.2.17) is a simpler equation than (6.2.11), and all of m, k, and P have been eliminated, except in as far as m and k appear in the ratio ω/ω_n.

Equation (6.2.17) is not the same as the nondimensionalized equation (6.2.8) for the mass–spring system with no external force, because time has been made dimensionless in a different way.

The choice of nondimensionalization in (6.2.12) and (6.2.12) at first appears arbitrary, but it is not. Suppose that the frequency of the external force were very low. Inertia would have no effect, and the spring would extend slowly as the force increased, reach a maximum extension P/k when πt reached $\pi/2$, decrease to 0 when πt reached π, decrease further to $-P/k$ when πt reached $3\pi/2$, and increase again to 0 when πt reached 2π. x is therefore the ratio of the displacement X to the maximum displacement that occurs if a force P is applied slowly. However, this is not to say that the nondimensionalization in (6.2.12) and (6.2.13) is the only possible one. Dimensional analysis is inherently creative and exploratory, and there is no set recipe for finding the nondimensionalization that will turn out to be the most useful and illuminating. Constructive play — we should not apologize for the word 'play' — with different possibilities often leads to illuminating results.

ω_n defined by (6.2.18) is the natural frequency for the mass–spring system alone, as we can see from (6.2.10). The ratio ω/ω_n is the ratio between the driving frequency ω of the external force and the natural frequency ω_n of the system.

It follows again that if we can find a solution of (6.2.17) for a particular value of ω/ω_n, then that solution is generally applicable to any spring–mass system driven by a sinusoidally varying external force with the same ratio ω/ω_n. If we can carry out solutions for a range of values of ω/ω_n, we have a universal result that we can apply to any system.

In this instance the solution is again straightforward. Thinking of the system physically, we expect an oscillating force with a frequency ω to induce an oscillatory response with the same frequency, and so we look for a response of the form

$$X = B \sin \omega T$$
$$x = b \sin t \qquad\qquad (6.2.19)$$

Substituting into (6.2.17)

$$-\left(\frac{\omega}{\omega_n}\right)^2 b \sin t = -b \sin t + \sin t \tag{6.2.20}$$

for all t, and so, dividing by $\sin t$ and rearranging

$$b = \frac{1}{1 - \left(\frac{\omega}{\omega_n}\right)^2} \tag{6.2.21}$$

If ω/ω_n is small compared to 1, the second term in the denominator is negligible compared to 1, and b is 1: this is the quasistatic response when the driving frequency is much smaller than the natural frequency. If ω/ω_n increases toward 1, the denominator gets smaller, and b increases. If ω/ω_n is exactly 1, the denominator is 0, and b is infinite: this corresponds to resonance between the natural frequency and the frequency of the driving force. If ω/ω_n is greater than 1, b is negative, so that the instantaneous displacement of the mass is in the opposite direction to the instantaneous value of the driving force. Finally, if ω/ω_n is very large, the denominator is large and negative, and b increases toward 0.

Again the system can be generalized. Imagine that we add to the system of Fig. 6.2 a damping device that produces a resistance to motion, to get Fig. 6.3. The damping device is conventionally represented by a dashpot (an idealized representation of a car shock-absorber). The force exerted on the mass by the dashpot is $-C\frac{dX}{dT}$ to the right, so that if the velocity is to the right ($\frac{dX}{dT} > 0$) the force is to the left and if the velocity is to the left ($\frac{dX}{dT} < 0$) the force is to the right.

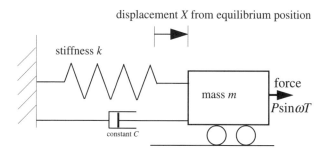

displacement X from equilibrium position

stiffness k

mass m

force
$P\sin\omega T$

constant C

Fig. 6.3. Mass–spring–dashpot system.

The equation of motion for the mass becomes

$$m\frac{d^2X}{dT^2} = -kX - C\frac{dX}{dt} + P\sin\omega T \tag{6.2.22}$$

Again change X and T to dimensionless variables x and t, following the scheme of (6.2.12) to (6.2.15). In the dimensionless variables, (6.2.22) becomes

$$m\frac{d^2\left(\frac{Px}{k}\right)}{d\left(\frac{t}{\omega}\right)^2} = -k\left(\frac{Px}{k}\right) - C\frac{d\left(\frac{Px}{k}\right)}{d(\frac{t}{\omega})} + P\sin t \tag{6.2.23}$$

which simplifies to

$$\left(\frac{\omega}{\omega_n}\right)^2\frac{d^2x}{dt^2} = -x - 2\zeta\left(\frac{\omega}{\omega_n}\right)\frac{dx}{dt} + \sin t \tag{6.2.24}$$

where

$$\zeta = \frac{C}{2\sqrt{mk}} \tag{6.2.25}$$

and ω_n is $\sqrt{k/m}$ as in Eq. (6.2.18).

Equation (6.2.24) is the governing differential equation, but now with two dimensionless ratios, ω/ω_0 and $C/\sqrt{(mk)}$. This equation does not have solutions of the form (6.2.9), because of the presence of the dx/dt term. Instead we look for a broader class of solutions

$$\begin{aligned} X &= A\cos\omega T + B\sin\omega T \\ x &= a\cos t + b\sin t \end{aligned} \tag{6.2.26}$$

Substituting into (6.2.24)

$$-\left(\frac{\omega}{\omega_n}\right)^2(a\cos t + b\sin t) = -(a\cos t + b\sin t)$$

$$-2\zeta\left(\frac{\omega}{\omega_n}\right)(-a\sin t + b\cos t) + \sin t \tag{6.2.27}$$

and, since this must hold for any value of t, the coefficients of $\cos t$ and $\sin t$ must both be 0, so that

$$0 = b \left(\frac{\omega}{\omega_n} \right)^2 - b + 2a\zeta \left(\frac{\omega}{\omega_n} \right) + 1 \tag{6.2.28}$$

$$0 = a \left(\frac{\omega}{\omega_n} \right)^2 - a - 2b\zeta \left(\frac{\omega}{\omega_n} \right) \tag{6.2.29}$$

simultaneous equations in a and b, whose solutions are

$$a = -\frac{2\zeta \left(\frac{\omega}{\omega_n} \right)}{\left(\frac{\omega}{\omega_n} \right)^4 - (2 - 4\zeta^2) \left(\frac{\omega}{\omega_n} \right)^2 + 1} \tag{6.2.30}$$

$$b = -\frac{\left(\frac{\omega}{\omega_n} \right)^2 - 1}{\left(\frac{\omega}{\omega_n} \right)^4 - (2 - 4\zeta^2) \left(\frac{\omega}{\omega_n} \right)^2 + 1} \tag{6.2.31}$$

If c is 0, a is 0 and b is given by (6.2.21) as before. The response (6.2.26) is sinusoidal, but now the displacement is not in phase with the driving force. The amplitude of the response is $\sqrt{(a^2 + b^2)}$, which is

$$\sqrt{a^2 + b^2} = \frac{1}{\sqrt{\left(1 - \left(\frac{\omega}{\omega_0} \right)^2 \right)^2 + \left(2\zeta \frac{\omega}{\omega_0} \right)^2}} \tag{6.2.32}$$

and the phase angle by which the displacement response lags the force is

$$\arctan \left(\frac{a}{b} \right) = \frac{2\zeta \frac{\omega}{\omega_n}}{1 - \left(\frac{\omega}{\omega_n} \right)^2} \tag{6.2.33}$$

These are now universal results that can be applied to any linear system. Like the governing differential equation, they involve just the two dimensionless ratios, ω/ω_n and the damping ratio ζ. Figure 6.4 is a chart that plots the response amplitude $\sqrt{(a^2 + b^2)}$ and phase angle ϕ against ω/ω_n for different values of ζ.

The advantage of carrying out the above analysis is that all the results for linear mass–spring–dashpot systems subject to sinusoidal forces can be summarized in two charts.

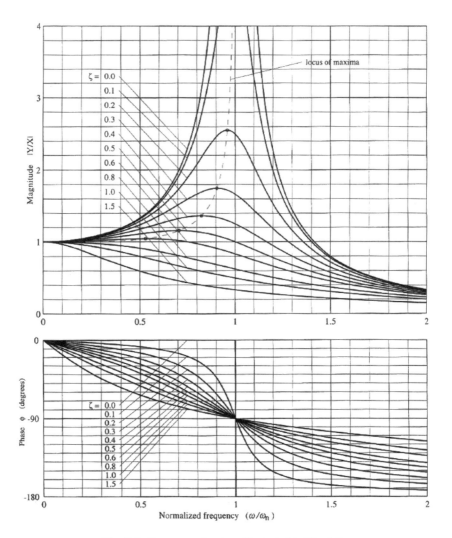

Fig. 6.4. *Response of system illustrated in Fig. 6.3.*

6.3   Laying underwater pipelines

Underwater pipelines are generally laid on the seabed by the scheme sketched in Fig. 6.5. The construction is based on a moored or dynamically positioned barge. The pipeline is built on a ramp along the length of the barge. Lengths of pipe are brought from shore and lined up at the upper end of the ramp, and then pass through a series of welding stations as the barge moves forward. Tensioners apply a tension force to the pipe near the stern end of the ramp. The pipe leaves the barge at the stern, and its configuration immediately beyond the stern is a convex-upward curved section, called the overbend, where it is supported on rollers by a structure called stinger. The stinger is a substantial structure, often nearly 100 m long, generally constructed as a single open framework rigidly fixed to the barge, but it sometimes has one or more buoyant segments hinged to each other and to the barge.

The pipe loses contact with the stinger at the lift-off point just above the end. It continues downward through the water as a long suspended span, a concave-upward curve called the "sagbend," and reaches the seabed tangentially at the touch-down point.

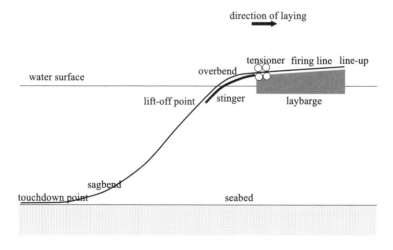

Fig. 6.5. Underwater pipelaying.

The shape taken up by the pipe in the sagbend is primarily controlled by the interaction between the applied tension and the submerged weight of the pipe, and to a lesser extent by the flexural rigidity of the pipe. If the applied tension is increased, the curvature of the pipe in the sagbend decreases, so that the sagbend becomes longer and flatter, while the touchdown point moves further from the barge and the liftoff point moves up the stinger. If the applied tension is reduced, the sagbend curvature increases, and if the tension is reduced too far the bending curvature may become excessive and the pipe may buckle. On the other hand, the shape taken up by the pipe in the overbend is controlled by the stinger geometry: tension has almost no effect on the overbend configuration if the stinger is rigid, and only a small effect if the stinger is segmented. The stinger has to be quite long, because otherwise the stinger curvature will not match the suspended-span curvature and excessive bending will occur at the end of the stinger. The technology is described in greater detail by Palmer and King.[1]

Marine pipeline construction in water depths between 100 and 5000 m is inevitably on a large physical scale that requires great lengths of pipe. On that large-scale, even large-diameter pipe is extremely flexible. The engineer can take advantage of this flexibility to bend and twist the pipe into complex three-dimensional forms, in order to place the pipe where he/she wants it to rest on the seabed, and to align it for connection to fixed and floating systems. He/she will want to know the configuration that the pipe will take up, to determine the forces that need to be applied and the stresses that will be induced, and to optimize the positioning strategy.

It turns out to be instructive to derive the governing equations and to write them in terms of dimensionless variables. Several useful results then follow.

First, we examine the similarity conditions. The strategy is to derive the governing equations for the two-dimensional case, in which the suspended pipeline lies in a vertical plane, the pipe is being laid on the bottom along a line in the same plane, and there are no transverse currents (perpendicular to the plane of Fig. 6.5).

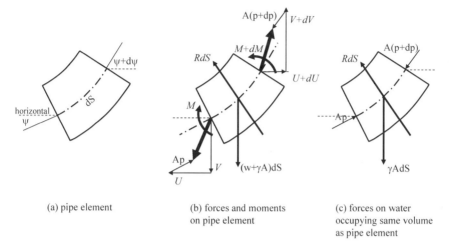

(a) pipe element

(b) forces and moments
on pipe element

(c) forces on water
occupying same volume
as pipe element

Fig. 6.6. Element of submerged pipeline.

Figure 6.6a shows an element of a pipe, and Fig. 6.6b the forces and moments that act on it. Distance along the pipe is denoted S and inclination of the axis to the horizontal is ψ. The pipe cross-sectional area is A (uniform along the length), the submerged weight is w per unit length, and the unit weight of water is ρg, so that the total weight per unit length is $w + \rho g A$. The force at each end of the element is represented by a heavy arrow, and can be divided into three components, a force in the local direction of the pipe axis equal to A multiplied by the local hydrostatic pressure p, an additional horizontal component U, and an additional vertical component V.

The hydrostatic force on the curved outer surface of the pipe element (but not the ends) is R per unit length. It can most easily be determined by imagining the element of pipe replaced by an element of water occupying the same volume. Figure 6.6c shows that element of water and the forces that act on it.

The element of pipe in Fig. 6.6b is in equilibrium. Figure 6.6c shows an element of water that occupies the same space as the element of pipe in Fig. 6.6b. That element too is in equilibrium. It follows that the differences between the forces in Fig. 6.6b and the forces in Fig. 6.6c must be in equilibrium. The corresponding equilibrium equations are simply:

$$\frac{dU}{dS} = 0 \tag{6.3.1}$$

$$\frac{dV}{dS} = w \tag{6.3.2}$$

$$\frac{dM}{dS} = U \sin \psi - V \cos \psi \tag{6.3.3}$$

In addition, we assume that the pipe would be perfectly straight if it were unloaded, and that it responds elastically in bending. The curvature $d\psi/dS$ is proportional to the moment M, and

$$M = F\frac{d\psi}{dS} \tag{6.3.4}$$

where F is the flexural rigidity.

Equation (6.3.1) tells us that U is independent of S and that U must be constant along the length of the suspended span. Substituting from (6.3.4) into (6.3.3), multiplying it by $\sec \psi$ and then differentiating with respect to S, and finally eliminating dV/ds by using (6.3.2), the governing equation that determines the configuration of the pipe is[1,2]

$$F\frac{d}{dS}\left(\sec \psi \frac{d^2\psi}{dS^2}\right) = U \sec^2 \psi \frac{d\psi}{dS} - w \tag{6.3.5}$$

Now nondimensionalize lengths such as S with respect to an arbitrary length H, so that

$$s = S/H \tag{6.3.6}$$

and in terms of this new variable (6.3.5) becomes

$$\frac{F}{wH^3}\frac{d}{ds}\left(\sec \psi \frac{d^2\psi}{ds^2}\right) = \frac{U}{wH} \sec^2 \psi \frac{d\psi}{ds} - 1 \tag{6.3.7}$$

The notion of changing the variables to make them dimensionless can be taken further. In (6.3.6) H was an arbitrary length. If now we set

$$H = \frac{U}{w} \tag{6.3.8}$$

the governing Eq. (6.3.7) becomes

$$\frac{Fw^2}{U^3}\frac{d}{ds}\left(\sec\psi\frac{d^2\psi}{ds^2}\right) = \sec^2\psi\frac{d\psi}{ds} - 1 \tag{6.3.9}$$

which now has only one dimensionless group.

The group could have been derived by the methods in Chap. 4. Taking the fundamental dimensions as M, L, and T, the dimensions of F, w, and U are

$$[F] \quad [ML^3T^{-2}]$$
$$[w] \quad [MT^{-2}]$$
$$[U] \quad [MLT^{-2}]$$

There are three variables and three fundamental dimensions. However, the dimension matrix $\begin{bmatrix} 1 & 1 & 1 \\ 3 & 0 & 1 \\ -2 & -2 & -2 \end{bmatrix}$ is singular, as can be seen by adding twice the first row to the third row, and its rank is 2 rather than 3. There is $3 - 2 = 1$ independent dimensionless group, and it is Fw^2/U^3: that it is the only dimensionless group can be checked by the methods described in Chap. 3.

Over most of the suspended span the Fw^2/U^3 term then has a negligibly small effect, but it becomes important at the lift-off point where the pipe lifts off the stinger, and at the touchdown point where the pipe reaches the seabed. If we idealize the seabed as rigid, the section of the pipe beyond the touchdown point (to the left in Fig. 6.5) is straight. The curvature $d\psi/ds$ is 0 immediately to the left of the touchdown point. If we ignore the Fw^2/U^3 term in (6.3.9), the curvature immediately to the right of the touchdown point is 1 (from (6.3.11)). A solution that ignores the Fw^2/U^3 term implies a sudden jump discontinuity in $d\psi/ds$ at the touchdown point, from 1 in the suspended span to 0 on the bottom, and that implies a sudden change in bending moment. Since there can be no externally applied moment at the touchdown point, that solution breaks down. The problem is resolved by incorporating into the solution narrow boundary layers that match the boundary conditions at liftoff and touchdown.[2]

Section 7.2 describes a useful physical model of pipelaying based on Eq. (6.3.7).

6.4 Partial differential equations describing fluid flow

An incompressible fluid in motion will be influenced by its inertia, its viscosity, pressures within it, and by boundaries with solid objects or with other fluids. The governing equations are derived by considering the forces that act on a small element of the fluid, and equating the resultant force on the element to its acceleration multiplied by its mass.[3] Referred to Cartesian axes x, y, and z (where z is vertical and positive upwards), and denoting the corresponding components of velocity as u, v, and w, pressure p, density ρ, and viscosity μ, the equations are

$$\rho\left(\frac{\partial u}{\partial t} + u\frac{\partial u}{\partial x} + v\frac{\partial u}{\partial y} + w\frac{\partial u}{\partial z}\right) = -\frac{\partial p}{\partial x} + \mu\left(\frac{\partial^2 u}{\partial x^2} + \frac{\partial^2 u}{\partial y^2} + \frac{\partial^2 u}{\partial z^2}\right)$$

$$\rho\left(\frac{\partial v}{\partial t} + u\frac{\partial v}{\partial x} + v\frac{\partial v}{\partial y} + w\frac{\partial v}{\partial z}\right) = -\frac{\partial p}{\partial y} + \mu\left(\frac{\partial^2 v}{\partial x^2} + \frac{\partial^2 v}{\partial y^2} + \frac{\partial^2 v}{\partial z^2}\right)$$

$$\rho\left(\frac{\partial w}{\partial t} + u\frac{\partial w}{\partial x} + v\frac{\partial w}{\partial y} + w\frac{\partial w}{\partial z}\right) = -\frac{\partial p}{\partial z} + \mu\left(\frac{\partial^2 w}{\partial x^2} + \frac{\partial^2 w}{\partial y^2} + \frac{\partial^2 w}{\partial z^2}\right) - \rho g$$

$$(6.4.1)$$

called the Navier–Stokes equations. The four terms in the large bracket on the left-hand side add up to the acceleration. The first term represents the change of velocity with time at a fixed location, and the other three the convective acceleration as the fluid moves to locations where the velocity is different. In addition

$$\frac{\partial u}{\partial x} + \frac{\partial v}{\partial y} + \frac{\partial w}{\partial z} = 0 \qquad (6.4.2)$$

represents the incompressibility condition. The equations can be written more concisely in vector notation or in Cartesian tensor notation.[3,4]

It is instructive to write the equations in terms of dimensionless variables. Choose a characteristic length L and a characteristic velocity U, so that L/U is a characteristic time. Create dimensionless variables, identified by a prime ', by nondimensionalizing lengths such as x with respect to L, velocities such as u with respect to U, and time with respect to L/U, so that

$$x' = \frac{x}{L} \quad \text{etc.}$$

$$u' = \frac{u}{U} \quad \text{etc.} \tag{6.4.3}$$

$$t' = \frac{t}{L/U}$$

and in addition nondimensionalize p with respect to ρU^2, so that

$$p' = \frac{p}{\rho U^2} \tag{6.4.4}$$

In terms of the primed variables (6.4.1) become

$$\frac{\partial u'}{\partial t'} + u'\frac{\partial u'}{\partial x'} + v'\frac{\partial u'}{\partial y'} + w'\frac{\partial u'}{\partial z'} = -\frac{\partial p'}{\partial x'} + \frac{\mu}{\rho U L}\left(\frac{\partial^2 u'}{\partial x'^2} + \frac{\partial^2 u'}{\partial y'^2} + \frac{\partial^2 u'}{\partial z'^2}\right)$$

$$\frac{\partial v'}{\partial t'} + u'\frac{\partial v'}{\partial x'} + v'\frac{\partial v'}{\partial y'} + w'\frac{\partial v'}{\partial z'} = -\frac{\partial p'}{\partial y'} + \frac{\mu}{\rho U L}\left(\frac{\partial^2 v'}{\partial x'^2} + \frac{\partial^2 v'}{\partial y'^2} + \frac{\partial^2 v'}{\partial z'^2}\right)$$

$$\frac{\partial w'}{\partial t'} + u'\frac{\partial w'}{\partial x'} + v'\frac{\partial w'}{\partial y'} + w'\frac{\partial w'}{\partial z'} = -\frac{\partial p'}{\partial z'} + \frac{\mu}{\rho U L}\left(\frac{\partial^2 w'}{\partial x'^2} + \frac{\partial^2 w'}{\partial y'^2} + \frac{\partial^2 w'}{\partial z'^2}\right)$$
$$-\frac{gL}{U^2}$$

$$\tag{6.4.5}$$

which contain two dimensionless groups, $\mu/\rho U L$ the reciprocal of the Reynolds number Re defined in Chap. 4, and gL/U^2 the reciprocal of the square of the Froude number Fr defined in Sec. 3.5 and later discussed in Chap. 4. Equation (6.4.2) becomes

$$\frac{\partial u'}{\partial x'} + \frac{\partial v'}{\partial y'} + \frac{\partial w'}{\partial z'} = 0 \tag{6.4.6}$$

It follows that if we can solve (6.4.5) and (6.4.6) for particular values of the parameters, the geometry and the boundary conditions, we have a family of solutions that will apply generally, provided that Re and Fr take the same values.

A comparison of (6.4.1) and (6.4.5) shows that the Fr term in (6.4.5) comes from the $-\rho g$ term in (6.4.1). If the problem we are interested in has no free surface, we can eliminate the need for Fr similarity by splitting the pressure p into two parts, a hydrostatic component $-\rho g z$ induced by the weight of the fluid, and a remainder $p + \rho g z$ induced by the flow dynamics.

This argument confirms the observation in Sec. 6.3 that dimensionless groups often turn up naturally if we write governing equations in dimensionless form.

6.5 Problems

6.1   Figure 6.7a is a generalization of the system considered in Sec. 6.2. Now there are two masses m, connected together by a spring with stiffness k and each connected to a rigid foundation by a spring with the same stiffness. Find the two differential equations that govern the motions of this system, transform them into nondimensional forms, and use the results to find the two natural frequencies of the system.

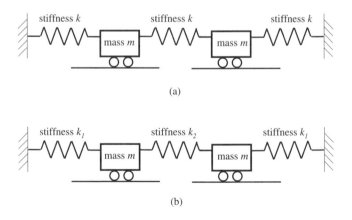

(a)

(b)

Fig. 6.7. Mass–spring systems.

6.2 Figure 6.7b is a further generalization. Now the center spring has stiffness k_2 and the end springs each have stiffness k_1. Carry out the analysis for this case, and consider what happens when $k_1 \gg k_2$ and when $k_1 \ll k_2$.

6.3 A floating ice sheet can carry vertical loads. Under additional external loads, the ice sheet deflects and bends downward, so that, in the area around the load, the local water pressure applied to the undersurface of the ice is increased.

Consider a sheet of ice with uniform thickness t, floating on still water with unit weight ρg per unit volume, and suppose the sheet to remain elastic, with elastic modulus E and Poisson's ratio v. x and y are orthogonal reference axes in the plane of the ice sheet, and w is vertical deflection from an initial unloaded position, positive downward. Plate theory[5] shows that w is governed by the fourth-order partial differential equation

$$D\left(\frac{\partial^4 w}{\partial x^4} + 2\frac{\partial^4 w}{\partial x^2 \partial y^2} + \frac{\partial^4 w}{\partial y^4}\right) = \rho g w \tag{6.5.1}$$

where $D = \frac{Et^3}{12(1-v^2)}$.

(i) Find a nondimensionalization of (6.5.1) that turns it into a form in which neither D nor ρg appears explicitly.

(ii) The tensile strain in the x-direction on the undersurface of the ice is given by

$$\varepsilon_{xx} = (t/2)\frac{\partial^2 w}{\partial x^2} \tag{6.5.2}$$

A simple (and oversimplified) approach to fracture says that the ice sheet will crack when the maximum value of the tensile strain reaches a critical value. Suppose the ice to carry a single concentrated load P. Use the equations to find how the maximum tensile strain depends on P and t.

(iii) A simple rule of thumb in Arctic engineering practice says that the maximum load in pounds that sea ice can carry safely is 50 times the square of the ice thickness measured in inches. Is this consistent with your result in (ii)? (Do not use this rule on its own to decide that you can walk on thin ice: ice is highly variable and often cracked).

(iv) If the load is spread out, the ice can carry more load. Use your results above to derive a simple rule that tells an engineer roughly how far the load must be spread out before the safe load is significantly increased.

6.4 Equations (6.4.1) are the equations of motion referred to fixed reference axes, which are stationary or move with uniform velocity. If however the reference axes are rotating, the acceleration has an additional component, which accounts for the fact that an object moving with a uniform velocity measured with respect to rotating axes nevertheless has an acceleration. That acceleration is called the Coriolis acceleration, after Gaspard Gustave de Coriolis (1792–1843). Rewriting (6.4.1) in vector notation, but after removing the viscosity and gravitation terms, we have

$$\rho \left(\frac{\partial \mathbf{u}}{\partial t} + \mathbf{u}\nabla\mathbf{u} + 2\Omega \times \mathbf{u} \right) = \nabla p \tag{6.5.3}$$

where variables marked in bold are vectors. The new term $2\Omega \times \mathbf{u}$ is the Coriolis acceleration, and Ω is the angular velocity vector describing the rotation of the frame of reference. Ω can usefully be thought of as $\Omega\mathbf{e}$, a magnitude Ω multiplied by a unit vector $\mathbf{e}$ in the direction of the axis of rotation.

Nondimensionalize velocities with respect to a characteristic velocity U, lengths with respect to a characteristic length L, pressure with respect to ρU^2, and therefore times with respect to L/U. Show that the nondimensionalized form of (6.5.3) then includes a dimensionless group $U/L\Omega$, called Rossby number after the Swedish meteorologist Carl-Gustaf Rossby (1898–1957) who strengthened the link between meteorology and fluid dynamics. The Coriolis acceleration is important when Rossby number is small, as it is for large-scale atmospheric and oceanographic motions on the Earth, which is of course rotating. It is almost always insignificant on a small scale. (It is a myth that water draining out of a washbasin rotates clockwise or counter-clockwise because of the Earth's rotation.)

References

1. Palmer, AC and RA King (2004). *Subsea Pipeline Engineering*. Tulsa, OK: Pennwell.

2. Palmer, AC, G Hutchinson and JE Ells (1974). Configuration of submarine pipelines during laying operations. *American Society of Mechanical Engineers, Journal of Engineering for Industry*, 96, 1112–1118.

3. Batchelor, GK (1967). *An Introduction to Fluid Dynamics*. Cambridge: Cambridge University Press.

4. Prager, W (1956). *Introduction to Mechanics of Continua*. Boston: Ginn.

5. Timoshenko, SP and JM Gere (1936). *Theory of Elastic Stability*. New York: McGraw-Hill.

Chapter 7

Physical Models

7.1 Introduction

Physical models are an attractive way of learning about the behavior of engineering systems. They can be constructed easily and relatively cheaply, and different options can be tried out rapidly. They appeal to unconscious happy childhood memories of launching model ships onto ponds, playing with dolls' houses, making sandcastles and model aircraft, and damming streams. Models are fun to work with: there is nothing wrong with that!

But models can be deeply misleading. The late Professor John Baker, the father of plastic design of steel structures, once said to the author that in his experience "nothing has so much misled structural engineers as models".[1] The point was further developed by Svendsen.[2] He was writing in the context of models of hydrodynamic systems that include water waves, but his wise remarks are much more widely applicable:

> *"To me there seem to be three virtually complementary goals that one may pursue with a physical model.*
>
> *(a) One is to seek qualitative insight into a phenomenon not yet described or understood.*
> *(b) Another is to get measurements to verify (or disprove) a theoretical result.*
> *(c) The third is to obtain measurements for phenomena so complicated that so far they have not been accessible for theoretical approaches.*
>
> *As an introduction it seems worthwhile to discuss briefly some potentials for controversies in the interpretation of results from*

a physical model. The point I have in mind is particularly relevant for models of type (b) and may be expressed by the following:

Theorem 1 *If there is a discrepancy between a theory and the experiment carried out to verify it, it is likely to be due to inaccuracies in the experiment.*

Stated in this (slightly provocative) form, the theorem may be taken as a sign of overconfidence in the theoretical results. It should not. On the contrary, it is a warning against overconfidence in experimental results. This is because I believe in:

Theorem 2 *It is far more difficult to make good experiments than it is to make good theories.*

Whereas Theorem 1 may be proved statistically, Theorem 2 will always remain a personal point of view. However, I feel that Theorem 1 turns out to be correct often enough to convince me about Theorem 2.

Theorem 1 also becomes less provocative if one realizes that the type of experimental 'errors' seen in connection with attempts to check theoretical results are often simply that the experiments are carried out under conditions not corresponding to the basic assumptions of the theory...."

We can add the well-known remark that actually makes a slightly different point:

"Everyone believes an experiment, except the person who carried it out. Nobody believes a theory, except the person who developed it"

A problem often met is that some effects have a large influence in a model but almost no significance at full scale. Ocean waves, for example, are almost unaffected by surface tension (except locally in the formation of whitecaps and spray), but in a small-scale model surface tension has a significant effect. Ice is an extremely brittle material, with a fracture toughness one-tenth that of glass, and there is a ductile–brittle transition, so that ice-structure interactions in small-scale models are below the transition and dominated by ductile behavior, whereas at full scale brittle behavior dominates. The result is that small-scale models of ice acting on structures give misleadingly large forces.

Loosely, small-scale models are too strong. The same thing happens in other kinds of structural engineering, among them ships and pipelines. Model structures are conveniently made of polymethyl-methacrylate (PMMA, 'Perspex,' 'Lucite') but PMMA creeps, so that at fixed loads its deformations increase with time, whereas most structural materials do not. Many other examples can be given. Section 7.4 examines some of the difficulties with physical models of estuaries.

Svendsen makes the strong case against naïve models, and too many models fall into that category. If some phenomenon is not properly understood, and a physical model is constructed, the problem is still not understood, and the experimenters may draw incorrect conclusions. The earlier chapters include examples, and in one instance it was rather more by luck than judgment that the model in fact gave useful and correct results. None of this ought to be interpreted as a blanket negative view that models can never be useful. In the words of Baker's colleague Jacques Heyman, "certain models are free of vice".[3] They can be revealing and useful, as long as we treat them with care and constructive scepticism. And we need to remember that theories too can be incorrect, and that if a problem is incompletely understood putting the calculations on a computer is likely to make things worse rather than better.

Over the past 40 years physical models have been progressively superseded by numerical methods. Many younger engineers are unfamiliar with models. They argue that anything that can be done with a model can be done better numerically on a computer, and that an application of models is an outmoded and unfashionable approach. They are wrong, and experience shows that models can be a more sophisticated and more efficient means of arriving at a good scheme than numerics are. Moreover, models have the advantage that they are necessarily three-dimensional, and that they compel engineers to think in the three dimensions of the real world, rather than having to accept the tyranny of the printout or the flat monitor screen.

Another advantage of models is that they are an excellent method of communicating with constructors, and with nonspecialists such as managers, regulators, bankers, and public-interest groups. Indeed, it is often a problem that a model makes things look easier than they actually are, so that nonspecialists come to feel that all difficulties have evaporated.

Let us look at some examples.

7.2   Structural models

Until the end of the eighteenth century, large structures were almost invariably built of stone or brick masonry. There were a few exceptions, such as mud multistory apartments in Yemen, suspension bridges in China made from bamboo and fiber ropes, and wooden churches in northern Europe, among them the spectacular timber space frame that supports the lantern in Ely Cathedral.

Masonry structures depend for their strength on the interaction between their weight and the geometrical arrangement of the blocks they are made from.[4] If they fail, they fail by rotations that open up hinges between the blocks, or by progressive consolidation and collapse of their soil foundations. The stresses in the blocks are very low compared to the compressive strength of the stone or brick, and crushing failures hardly ever occur.

Stability of a masonry structure can be examined by making a scale model from model blocks, with geometrically scaled blocks and a correct arrangement of the joints. The strength of the blocks is essentially irrelevant, and so the model blocks can be made from an easily worked material such as wood or plaster. Medieval architects made many models of this kind. In the context of ancillary structures needed during construction, Fitchen[5] wrote of them

"Their methods were eminently practical, pragmatic and direct. At least in the case of important falsework structures, the actual work was invariably laid out at full size, after the problems were all solved via accurate models. In fact, the design of every aspect and detail seems unquestionably to have been worked out almost

exclusively in terms of three-dimensional models of all sorts, both for details and for ensembles"

and he quoted Waldrain[6]

"When Hugh Herland, Master Carpenter to Richard II, designed Westminster Hall roof, his textbooks and stress diagrams were his innumerable models, which as we know occupied so much space that rooms in the King's Palace had to be reserved for them"

With the coming of large-scale manufacture of iron and steel, it became possible to build much larger truss and frame constructions, among them bridges, towers and mills. Many engineers built models. The pattern that developed was that engineers and architects devised daring and innovative concepts than ran far ahead of the analytic methods of the time, and had to use scale models to resolve the problems. Fairbairn (1789–1874) made experiments in 1845 on large tubular box girders designed by Robert Stephenson for the Britannia and Conwy railway bridges, where the trains run inside the tubes. Preliminary experiments had shown that the girders would fail by buckling of the thin plates that formed the tubes, and Fairbairn's research was used to prepare design rules.[7] It had originally been thought that the tubes would have to be supported by suspension chains, and the towers at the ends of the spans had been built to the required height, but in the end the chains were left out, which accounts for the rather odd appearance of the bridges. Hossdorf's thoughtful book[8] gives examples of models from the work of many of the great engineers and architects of the early twentieth century, among them Nervi, Torroja, Gaudi, and Le Corbusier.

Pippard and Baker[9] discuss experimental structural analysis with models. One of their examples is the Dome of Discovery, constructed for the Festival of Britain in London in 1951, before computers were available. The 110 m diameter braced dome had the shape of a segment of a sphere, and was a lattice framework of arched ribs, supported around its edge by a ring beam, in turn supported on 24 bipods, which allowed radial but not tangential movements. The interconnections between the ribs, the ring beam and the bipods were much too complicated for it to be possible to calculate all the forces by statics alone: in the jargon of structural mechanics, the structure was highly redundant. The behavior of the ring beam was

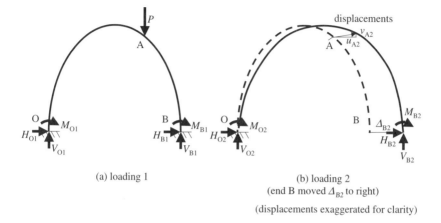

(a) loading 1

(b) loading 2
(end B moved Δ_{B2} to right)

(displacements exaggerated for clarity)

Fig. 7.1. Arch.

found from experiments on steel and plastic rings 457 mm across. A 3.2 m diameter model of the whole dome was used to find the forces in the lattice.

A useful theorem in model structural analysis is the Maxwell–Betti reciprocal theorem,[9] which applies to linear elastic structures, that is, to structures in which the stress and strain in the structure are linearly proportional to each other, so that yield, fracture, and creep do not occur. Imagine two different loadings on the structure, load 1 and load 2, and the corresponding displacements of the points at which the loads are applied. 'Corresponding' means at the same point and in the same direction, so that if load 1 include a horizontal external force U_{1A} (positive to the right) at point A, the corresponding displacement u_{1A} is the horizontal displacement at A (positive to the right). Similarly, if the loading includes a vertical force V_A at point A (positive upward), its corresponding displacement v_{1A} is the vertical displacement at A (positive upward). If the loading includes a clockwise external moment M_{1A}, its corresponding displacement ϕ_{1A} is the clockwise rotation at A. The theorem says that if all the loads in loading 1 are multiplied by the corresponding displacements under loading 2, and added up over the whole structure, then the result is equal to that got by multiplying the loads in loading 2 by the corresponding displacements in loading 1, and again added up over the whole structure.

Imagine then an arch OAB built in at either end (Fig. 7.1a), and on it a single external vertical load P applied at point A: call that loading 1. We wish to find the horizontal force H_{1B} that that load P at A induces at the right-hand end B, and the moment induced at the same point. Remove the load P, and move the right-hand end of the arch B a distance Δ_{B2} to the right, without letting B either move vertically or rotate (Fig. 7.1b): call that loading 2 (obviously different from loading 1). In particular, loading 2 induces at point A a vertical displacement v_{A2}, a horizontal displacement u_{A2}, and a rotation ϕ_{A2}. Now apply the theorem to loadings 1 and 2. Equating the two sums:

$$0 + (0)u_{A2} + (-P)v_{A2} + (0)\phi_{A2} + H_{B1}\Delta_{B2} + V_{B1}(0) + M_{B1}(0) = 0 + 0 + 0$$

point O   point A   point B   points O, A, and B

$$(7.2.1)$$

which simplifies to just

$$H_{B1} = \frac{v_{A2}}{\Delta_{B2}}P \tag{7.2.2}$$

The forces and moments in arches depend on the distribution of flexural rigidity and to a lesser extent on axial stiffness. If we make a model of the arch, we can carry out the experiment that corresponds to loading 2 on the model arch, measure the imposed displacement Δ_{B2} and the displacement u_{2A} at the load point A, and use (7.2.2) to determine H_{B1}, because if we have modeled the distribution of stiffness correctly the same ratio will apply to the model and to the prototype. Often axial stiffness will be insignificant, and we only have to model the distribution of flexural stiffness. Further examples are given in Refs. 9 and 10.

A third structural model example is the pipelaying problem described in Sec. 6.3; the reader should read that section first.

Equation (6.3.7) contains two dimensionless groups, U/wH the dimensionless horizontal tension and F/wH^3 the dimensionless flexural rigidity. It follows that if a model is constructed in such a way that those two groups are the same in the model as in the prototype, the configuration taken up by the model will be correctly scaled, in the sense that the relationship between ψ and s in the prototype will be the same as the relationship between ψ and s in the model. Denoting prototype and model quantities by subscripts p and m, the length scale factor is

$$\frac{H_p}{H_m} = \left(\frac{F_p/w_p}{F_m/w_m}\right)^{1/3} \tag{7.2.3}$$

All forces scale in the same way as U, and the force scale factor is

$$\frac{U_p}{U_m} = \left(\frac{F_p}{F_m}\right)^{1/3} \left(\frac{w_p}{w_m}\right)^{2/3} \tag{7.2.4}$$

The same similarity conditions could alternatively be derived by dimensional analysis, but the advantage of working by way of the governing equation is that it becomes clear why U is constant, which is not immediately obvious.

The above analysis is for the two-dimensional case. It extends to three dimensions, but that raises the possibility of torsion. The relevant additional dimensionless group is GJ/F, the ratio of the torsional rigidity to the flexural rigidity, where G is the shear modulus and for an axially symmetric cross-section J is the polar second moment of area. In the elastic range F is Young's modulus E multiplied by the second moment of area I about a diameter. For an axially symmetric section J/I is 2. For an isotropic elastic material E/G is $2(1 + \nu)$ where ν is Poisson's ratio. It follows that if a steel pipe is modeled, all that is required to take account of torsion is that Poisson's ratio be the same in the model as in the prototype. Concrete weight coating has a relatively minor effect on flexural and torsional stiffness (because concrete has much lower elastic moduli than steel, and cracks), but it may sometimes need to be taken into account. In most operations torsional moments are in reality small, because the pipe is free to rotate about its axis.

There is no need for the model pipe to be under water, provided that w_m is taken to be the air weight per unit length of the model, and w_p the submerged weight per unit length of the prototype. Moreover, there is no need for the model of the pipe itself to be a pipe, or to have the same diameter/thickness ratio as the prototype pipe: it can equally well be a wire or a rod. These are immense practical advantages.

The scale factors turn out to be convenient for models at a sensible scale that can be economically built in an engineering office, so that engineers devising construction scheme can routinely test their ideas on the model. The first model of this type was built in 1972, as part of BP preparations for the laying of the Forties 1 pipeline in the North Sea.[11] The prototype pipeline had a 812.8 mm (32-in.) outside diameter, 19.05 mm wall thickness, and a 1.1 kN/m (112 kg/m, 75 lb/ft) submerged weight. It was modeled by a 1.57 mm outside diameter steel tube. The length scale factor was 109 and the force scale factor 1.224×10^6, so that the 128 m maximum water depth was represented in a model 1.17 m high and a laybarge tension of 350 kN was modeled by 0.286 N (29.2 g weight). Though a tube was used to model the pipeline, a wire would have done equally well: this was the first application of this kind of modeling, and some psychological factors came into play.

Physical models that apply this principle have been applied to many marine pipeline construction projects. Some of the applications are described by Palmer *et al.*,[11] Kaustinen[12] and Brown and Palmer.[13]

In deep water, the effect of the flexural rigidity F term on the left-hand side of Eq. (6.3.7) is negligible except in short boundary layers close to the touchdown point on the seabed,[11,13] the lift-off point where the pipeline leaves the stinger, and the points of application of concentrated loads such as buoyancy tanks. It is then often possible to drop the requirement for F/wH^3 similarity.[13]

7.3 Aerodynamic models

Early aeronautical engineers made extensive use of models. Benjamin Robins (1707–1751) realized the importance of air resistance in finding the range of guns, and propelled models on a rotating arm driven by a falling weight. George Cayley (1773–1857) made many models of gliders, some of them remarkably modern in appearance, and flew some as kites. Francis Wenham (1824–1908) worked in many fields, among them stereomicroscopy and photography, wrote one of the key papers in the development of aeronautics,[14] and was the first to use the word 'airplane.' He built the first wind tunnel in 1871. The Wright brothers mounted models on the front of a bicycle, and later constructed their own wind tunnel. Wind tunnels have a huge literature, and Bradshaw[15] is an introduction.

The first task is to find the lift and drag on aerofoil wings. If the relevant parameters are the velocity U of the aerofoil relative to the air, a length ℓ, the air density ρ, the air viscosity μ, and the lift F_L per unit length along the wing, then the analysis is essentially the same as for ships in Chap. 4. The dimensions are

velocity U	$[LT^{-1}]$
length ℓ	$[L]$
air density ρ	$[ML^{-3}]$
air viscosity μ	$[MLT^{-1}]$
lift force/unit length	$[MT^{-2}]$

A dimensionless lift coefficient $\frac{F_L}{(1/2)\rho U^2 \ell}$ is a function of Reynolds number $\frac{\rho U \ell}{\mu}$. Fluid dynamics conventionally puts the $\frac{1}{2}$ into the definitions of lift and drag coefficients.[16] ℓ is taken as the chord of the aerofoil cross-section, the distance from the leading edge to the trailing edge: that is a source of misunderstanding when comparisons are made between aerospace practice and other fields of fluid mechanics. A drag coefficient can be defined in the same way.

Think first of a wind tunnel that operates close to atmospheric pressure and temperature. Density ρ is proportional to pressure p and inversely proportional to absolute temperature θ. Viscosity μ is independent of p and for air approximately proportional to $\theta^{0.76}$,[17] and so (ρ/μ) is proportional to $p/\theta^{1.76}$. At the heights airplanes fly at, pressure and temperature are significantly lower than on the ground. Comparing a model at sea level with an aircraft flying at 5000 m height, and using the values of the International Standard Atmosphere,[16] p is 101.3 kPa on the ground and 54.0 kPa at 5000 m, θ is 288.16 °K on the ground and 255.66°K at 5000 m, and $(\rho/\mu)_{\text{model}}/(\rho/\mu)_{\text{prototype}}$ is 1.55. That tells us that to get Reynolds number the same in a model as the prototype the velocity U in the model must be much increased to compensate for the reduction of ℓ.

However, there is a complication because air is compressible. Define compressibility β as fractional change of volume per unit increase of pressure. Its dimension is [1/pressure] which is $[M^{-1}LT^2]$. Looking for a dimensionless group that includes U, β, and ρ but if possible excludes μ and ℓ, we find that $U(\rho\beta)^{1/2}$ is dimensionless (as in Problem 3.2). That group has a simple physical interpretation. $(\rho\beta)^{-1/2}$ is the speed of sound,[18] if we interpret β as the adiabatic compressibility (the compressibility if when the gas is compressed it does not have time to exchange heat with its surroundings). $U(\rho\beta)^{1/2}$ is therefore the ratio of the aircraft speed to the speed of sound. It is called Mach number Ma (but often M in the fluid dynamics literature), after Ernst Mach (1838–1916), an Austrian physicist, physiologist, and philosopher who carried out important research in many fields at the universities of Graz and Prague. Mach number is the only dimensionless group that people outside engineering and physics have heard of, and the Concorde supersonic airliner used to have in its passenger cabin a meter that displayed the current value of Mach number.

In air at 15 °C and atmospheric pressure, the velocity of sound is 340 m/s. Compressibility has almost no effect when Ma is smaller than 0.5, but becomes important as Ma approaches 1, and close to 1 the whole character of the flow changes. At one time it was thought that there was a 'sound barrier' that would make it difficult to fly at speeds faster than sound, but that turned out to be incorrect.

Ground vehicles such as Formula 1 racing cars are not large and operate at relatively low speeds, and so it is possible to test one-third or one-half scale models at the correct Reynolds number, at close to atmospheric pressure.

Aircraft, on the other hand, fly at higher speeds than cars and are much larger. If we want to find a way of having both the correct Reynolds number $\rho U\ell/\mu$ and not too high a Mach number, we can try to make (ρ/μ) much larger in the model than in the prototype. Going back to the fact that (ρ/μ) is proportional to $p/\theta^{1.76}$, we ought to increase the pressure p and reduce the temperature θ. Modern wind tunnels do this. The first high-pressure wind tunnel operated at 500 kPa (5 bars) and went into service in 1923, at the Langley Research Center in the United States. The European Transonic Windtunnel (ETW)[19] near Köln in Germany uses nitrogen as the working fluid, and by injecting liquid nitrogen lowers the flow temperature to −163 °C (110 °K) and increases the pressure to 450 kPa (4.5 bars). Compared to 15 °C (288.1 °K) and 1 bar, $p/\theta^{1.76}$ is increased in the ratio

$(4.5/1)/(110/288.1)^{1.76}$, which is 24.5; this does not take account of the small difference between air and nitrogen. The test section is 2.4 m wide, 2 m high, and 9 m long. Aircraft models can be tested in the ETW at the same Reynolds numbers as full-scale aircraft, and semi-span models can reach Re 8.5×10^7 at Ma 0.8 and Re 6×10^7 at Ma 1.3.

In seawater, the sound velocity is about 1500 m/s, and so for even a high-speed surface ship or a torpedo moving at 50 m/s (100 knots) Ma is less than 0.04 and compressibility effects are insignificant. Compressibility is significant in underwater explosions, where velocities are much larger.

An alternative to physical models is to carry out numerical calculations. Since its start 50 years ago, computational fluid dynamics (CFD) has competed with physical models. One experimental fluid dynamicist told the writer "Ever since I started doing fluid dynamics, people have predicted the end of experiments".[20] That has not happened, and models in wind tunnels continue to be used very widely. Though the number of wind tunnel facilities tends to decrease, existing and new facilities such as the ETW run for more hours than ever. Models in wind tunnels provide a more accurate prediction of drag than CFD can, and are used to 'anchor' CFD. CFD is best at designs that are not radically different from previous designs, and is useful to explore the effects of parameter variations. Models also remain valuable for flexible objects and blunt objects for which flow separation is important.

7.4   Hydraulic models

An estuary where a river flows into the sea is generally broad and shallow, typically 10 km across and 10 m deep. The geometry is complicated. It is important to understand how the flow of water is influenced by the river flow and the tide, because that will determine how sediment will move, which in turn determines how sandbanks that might obstruct navigation will build and erode with time, and because it will influence the movement of pollutants. Engineers also need to know how the flow will be influenced by constructions such as breakwaters, training walls, and new harbors.

The analysis of a possible physical model starts from the underlying differential equation that governs the flow of the water. Positions are defined with respect to axes x, y, and z fixed with respect to the Earth. Axes x and y are horizontal and parallel to the local east and north directions, and z is vertical, measured positive upwards from a fixed datum level. Velocity has components u, v, and w.

The Earth rotates, and therefore the axes rotate with it. It follows that an element of fluid that moves with a constant velocity relative to the axes does nevertheless have an acceleration, because the axes are turning. Figure 7.2 shows the reference axes, a velocity vector, the three components of the velocity, and the axis of rotation of the earth. The axis of rotation lies in the y, z plane, at an angle $(\pi/2)$-ϕ to the y-axis, where ϕ is the latitude. The component of acceleration that corresponds to the rotation of the reference axes is called the Coriolis acceleration (see also Problem 6.4), and is twice the vector product (cross product) of the velocity and the angular velocity vector, whose components are $(0, \Omega \sin \phi, \Omega \cos \phi)$, where Ω is the angular velocity of the Earth about its axis.

The differential equations that govern the motion are got by considering a rectangular parallelepiped element with its faces perpendicular to the axes and its sides dx, dy, and dz. The general equations of motion in

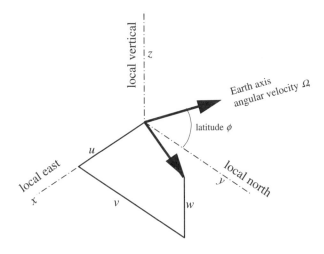

Fig. 7.2. *Reference axes.*

the three coordinate directions are[21]

$$\rho\left(\frac{\partial u}{\partial t} + u\frac{\partial u}{\partial x} + v\frac{\partial u}{\partial y} + w\frac{\partial u}{\partial z} - 2\Omega v \sin\phi\right) + \frac{\partial p}{\partial x} - \frac{\partial \sigma_{xz}}{\partial z} - \frac{\partial \sigma_{xy}}{\partial y} = 0$$

$$(7.4.1)$$

$$\rho\left(\frac{\partial v}{\partial t} + u\frac{\partial v}{\partial x} + v\frac{\partial v}{\partial y} + w\frac{\partial v}{\partial z} + 2\Omega u \sin\phi\right) + \frac{\partial p}{\partial y} - \frac{\partial \sigma_{zy}}{\partial z} - \frac{\partial \sigma_{xy}}{\partial x} = 0$$

$$(7.4.2)$$

$$\rho\left(\frac{\partial w}{\partial t} + u\frac{\partial w}{\partial x} + v\frac{\partial w}{\partial y} + w\frac{\partial u}{\partial z} - 2\Omega w \cos\phi\right) + \rho g + \frac{\partial p}{\partial z} - \frac{\partial \sigma_{xz}}{\partial x} - \frac{\partial \sigma_{yz}}{\partial y} = 0$$

$$(7.4.3)$$

The depth of an estuary is small by comparison with its length and breadth, and so the vertical components of velocity and acceleration are small by comparison with the horizontal components. In (7.4.3) the derivatives of the shear stresses s_{xz} and s_{yz} are negligible by comparison with ρg. Equation (7.4.3) therefore simplifies to

$$\rho g + \frac{\partial p}{\partial z} = 0 \qquad\qquad (7.4.4)$$

The water surface is at z_1, not necessarily constant because the surface is not horizontal. At the surface the pressure is atmospheric. On the scale of an estuary, variations in atmospheric pressure are generally negligible, though this might not be true for a broad estuary in an intense depression, and would not be true in the open sea. If the atmospheric pressure is everywhere p_a, integrating (7.4.4) from the surface to a height z

$$p = \rho g(z_1 - z) \qquad\qquad (7.4.5)$$

and so

$$\frac{\partial p}{\partial x} = \rho g \frac{\partial z_1}{\partial x} \qquad\qquad (7.4.6)$$

$$\frac{\partial p}{\partial y} = \rho g \frac{\partial z_1}{\partial y} \qquad\qquad (7.4.7)$$

and (7.4.1) and (7.4.2) become, after dividing through by ρ,

$$\frac{\partial u}{\partial t} + u\frac{\partial u}{\partial x} + v\frac{\partial u}{\partial y} - 2\Omega v \sin\phi + g\frac{\partial z_1}{\partial x} - \frac{1}{\rho}\frac{\partial \sigma_{xz}}{\partial z} - \frac{1}{\rho}\frac{\partial \sigma_{xy}}{\partial y} = 0 \qquad (7.4.8)$$

$$\frac{\partial v}{\partial t} + u\frac{\partial v}{\partial x} + v\frac{\partial v}{\partial y} + 2\Omega u \sin\phi + g\frac{\partial z_1}{\partial y} - \frac{1}{\rho}\frac{\partial \sigma_{zy}}{\partial z} - \frac{1}{\rho}\frac{\partial \sigma_{xy}}{\partial x} = 0 \qquad (7.4.9)$$

Consider now a scale model of an estuary. In the model we can simulate tides and rivers by controlling levels and flows at the boundary, and we can investigate how the flow change in response to changes introduced by civil engineering works, such as deepening channels by dredging, constructing new harbors, and adding training walls. A typical estuary is 10 km wide and 10 m deep, and so a model to a scale of 0.001 (1 to 1000) would be 10 m across and 10 mm deep. Ten millimerters is rather small: it would be difficult to construct an accurate model to such a small vertical scale, and a tidal range of 2 m would be scaled to only 2 mm. We therefore think of the possibility of a distorted model, in which the vertical scale is not necessarily the same as the horizontal scale, and ask ourselves if we can still satisfy the governing equations.

Horizontal lengths are scaled by a factor λ, so that x is replaced by λx and y by λy. Vertical lengths are scaled by a different factor ζ, so that z is replaced by ζz. Time is scaled by a factor τ, so that t is replaced by λt. Horizontal velocities, dimension $[LT^{-1}]$ are therefore scaled by a factor λ/τ, so that u is replaced by $(\lambda/\tau)u$. Angular velocity is scaled by ω, so Ω is replaced by $\omega\Omega$. Substituting into (7.4.8) and (7.4.9) and dividing through by λ/τ^2

$$\frac{\partial u}{\partial t} + u\frac{\partial u}{\partial x} + v\frac{\partial u}{\partial y} - 2\omega\tau\Omega v \sin\phi + \frac{g\tau^2\zeta}{\lambda^2}\frac{\partial z_1}{\partial x} - \frac{\tau^2}{\lambda\zeta\rho}\frac{\partial \sigma_{xz}}{\partial z} - \frac{\tau^2}{\lambda^2\rho}\frac{\partial \sigma_{xy}}{\partial y} = 0$$
$$(7.4.10)$$

$$\frac{\partial v}{\partial t} + u\frac{\partial v}{\partial x} + v\frac{\partial v}{\partial y} + 2\omega\tau\Omega u \sin\phi + \frac{g\tau^2\zeta}{\lambda^2}\frac{\partial z_1}{\partial y} - \frac{\tau^2}{\lambda\zeta\rho}\frac{\partial \sigma_{zy}}{\partial z} - \frac{\tau^2}{\lambda^2\rho}\frac{\partial \sigma_{xy}}{\partial x} = 0$$
$$(7.4.11)$$

First, considering the fifth term in each of the equations, similarity between the prototype and a model requires that the group $g\tau^2\zeta/\lambda^2 = 1$. The gravitational acceleration g has to be the same in the model as in the prototype, and so

$$\zeta = \left(\frac{\lambda}{\tau}\right)^2 \qquad (7.4.12)$$

and the vertical length scale ζ equals the square of the velocity scale λ/τ.

It is helpful to see what this implies. Think of a model with a horizontal scale λ equal to 0.001 (so that a 10 km broad estuary is represented by a model 10 m across) and a vertical scale ζ equal to 0.01 (so that an estuary depth of 10 m is modeled by a depth of 100 mm). From (7.4.11), the time scale τ is $\lambda/\sqrt{\zeta}$, which is 0.01, so that a 12 h 25 min period between one high tide and the next is simulated by 7.45 min, and a 1 knot (0.515 m/s) velocity in the prototype is simulated by $0.515(\lambda/\zeta)$ m/s, which is 0.0515 m/s. These seem sensible values for a large model.

Now consider the fourth term in (7.4.10) and (7.4.11), which represents the Coriolis acceleration. Similarity requires that $\omega\tau = 1$, if the model and the prototype are at the same latitude. This implies that the model must rotate faster than the prototype: with the numbers given earlier, the scaled rotational velocity must be 100 times the rotational velocity of the earth, which corresponds to one revolution every 1/100 of a day, which is 14.4 min. It is not straightforward to rotate a large model in this way, though it has occasionally been done. McDowell and O'Connor discuss this difficulty, and say that the failure to represent Coriolis accelerations may not matter if the bed is fixed, but that it is important if the bed is erodible. Coriolis accelerations are however invariably important in large-scale flows in seas and oceans away from the Equator (and in the atmosphere: it is because of the Coriolis effect that winds rotate anticlockwise around a depression in the northern hemisphere, but clockwise in the southern hemisphere).

Finally, consider the sixth and seventh terms in (7.4.10) and (7.4.11), which represent the effects of gradients of shear stress. Shear stress components like σ_{xz} that act on horizontal planes roughly parallel to the bed are much larger than components like σ_{xy} that act only on vertical planes, and therefore the seventh terms can be neglected. The sixth terms are not perfectly represented either, not least because Reynolds numbers are much smaller in the model. In modeling practice this is dealt with by making

Fig. 7.3. Model of outer estuary of Thames (photograph by courtesy of HR Wallingford).

the model seabed rougher than the prototype, and adjusting the roughness until the flows in the model match those observed.

Models of this kind were much used in the past. Figure 7.3 shows part of a model of the outer estuary of the Thames, constructed to investigate the effect of a proposed new airport on the Maplin Sands east of Foulness in England. However, physical models are expensive, and have various limitations, some of them discussed above, and most estuary modeling is nowadays computational.

Estuary models are only one kind of hydraulic model. Physical models are used in many other contexts, but a complete description would go beyond the scope of this book. Two more examples form Problems 7.4 to 7.7.

7.5 Geotechnical models

Imagine that we wish to investigate the design of a dam made of earth. One possibility would be to make the model out of the same soil as the prototype, but then the model would be misleadingly strong, because the relationship between the strength of the earth and the loads imposed on it by its own weight and the weight of the water would be wrong. If we take the soil to be characterized by a shear strength c and a unit weight ρg (the density ρ multiplied by the gravitational acceleration g), and describe the linear dimensions by the height H, the only dimensionless group is $\rho g H / c$, which is the Taylor number defined earlier. If the soil is the same in the model and the prototype, ρ and c are the same, and if g is also the same, $\rho g H / c$ is much smaller in a scaled-down model.

An alternative is to scale down c by mixing water into the soil. This is occasionally done, but it has several disadvantages. The model soil is extremely weak and difficult to work with, and its properties may change qualitatively, so that it becomes more like a thixotropic fluid than a soil.

Another alternative is use the same soil in the model as in the prototype, but to scale gravity up to compensate for the reduction in length scale. If Taylor number is the same in the model (m) as in the prototype (p)

$$\frac{\rho_m g_m H_m}{c_m} = \frac{\rho_p g_p H_p}{c_p} \tag{7.5.1}$$

and so if the soil is the same

$$\frac{g_m}{g_p} = \frac{H_p}{H_m} \tag{7.5.2}$$

and the ratio of gravity in the model to gravity in the prototype must be the reciprocal of the length scale.

This can be accomplished by rotating the model in a centrifuge. A 1/200 scale model, for example, is made from the prototype soil, mounted a centrifuge so that the radial direction in the model corresponds to the vertical direction in the prototype, and rotated to create an acceleration of 200 gravities. Centrifuge modeling was invented by Pokrovsky in Russia in the nineteen-thirties, picked up by Schofield in England in the nineteen-sixties, and recently widely developed in many countries. Many centrifuges are quite large: continuing the example, a 1/200 scale model of a prototype 200 m across in either horizontal direction is 1 m square in plan, and if it is 0.3 m deep it contains $0.3\,\text{m}^3$ of soil. If the density of the soil is

$2000 \, \text{kg/m}^3$, the model soil itself has a mass of $600 \, \text{kg}$, and at an acceleration of $1960 \, \text{m/s}^2$, 200 times the gravitational acceleration of $9.8 \, \text{m/s}^2$, the 'weight' of the model is $1.176 \, \text{MN}$, about 120 tonnes. If the centrifugal acceleration is not to vary more than 10% within the depth of the model, the radius to the mid-depth has to be 3 m. The rate of rotation is $25.6 \, \text{radians/s}$, which is 244 rpm. The centrifuge structure has to be robust, and it has to carry the forces induced by its own weight as well as the model weight.

The modeling principle described above assumes that the relevant parameters that describe the strength all have the dimensions of stress, $[F/L^2]$, where F is force. The strength of brittle materials such as ice and rock is governed by different parameters such as fracture toughness, whose dimension is $[F/L^{3/2}]$. The modeling law is then different,[22–24] and the acceleration in the centrifuge has to be the linear scale to the power $3/2$.

Scaling down the lengths in a model has the effect of speeding up diffusion processes, such as heat transfer and consolidation by water diffusion within the soil. In consolidation, for example, the relevant dimensionless group is $\kappa t / H^2$, where t is time and κ a water movement diffusivity that geotechnics calls coefficient of consolidation. Accordingly, if lengths are scaled down by a factor n, and the diffusivity is the same in the model as in the prototype (because the soil is the same), consolidation is speeded up by a factor n^2. In a $1/200$th scale model, a consolidation process that would take 10 years in the field takes in the model $(10 \text{ years})/200^2$, which is 2.2 h.

Centrifuge modeling is controversial, and has both passionate enthusiasts and detractors. Its long-term place in geotechnics remains uncertain. Even that statement is controversial.

7.6 Problems

7.1 A ship trying to anchor sometimes drags its anchor across the seabed. Severe damage occurs if the anchor hits an underwater pipeline. A pipeline is sometimes protected by placing a continuous heap of broken rock over it, but that is expensive, and so it is important not to place more rock than necessary.

A engineer wishes to model this interaction, and she/he is primarily concerned to find out how much rock is needed so that the anchor drags through the rock without touching the pipeline. Considering first the case where the anchor is dragging very slowly — so that dynamic effects to do with inertia are not important — find the relevant similarity conditions that a model must satisfy.

It would be convenient if the prototype interaction under water could be modeled correctly in a dry model in air rather than water. Is that possible?

Now suppose that the anchor is moving faster, so that dynamic effects may be important, and find the relevant similarity conditions.

7.2   The mouth of a fishing trawl net is held open by two large and heavy boards ('trawlboards,' 'otterboards'), one at each side and rigged to that they act like underwater kites. A typical board is 5 m long, 3 m high, and has a mass of several tonnes. If the board strikes an underwater pipeline, it can drag the pipeline sideways and damage the coating.

An engineer wishes to model this situation, and to find the time history of the force between the board and the pipeline. Find the similarity conditions that the model should satisfy. Assume first of all that the pipeline is fixed and undeformable. Then generalize the analysis to include the pipeline's weight and flexural stiffness. Stiffness is described by flexural rigidity, defined in Problem 4.2.

7.3   Wind can excite oscillations in a suspension bridge, and in extreme case the oscillations can bring the bridge down: the Tacoma Narrows bridge failure in 1940 is a famous example. This can be modeled by a model bridge in a wind tunnel. Derive the relevant similarity conditions, assuming for a start that

(i) The deck can be modeled as a continuous beam described by its span ℓ, mass per unit length m, flexural rigidity EI (ratio of bending moment to curvature, Problem 4.15), and torsional rigidity GJ (ratio of torque to twist per unit length;

(ii) The cables and hangers are infinitely flexible, and their mass is negligible by comparison with the mass of the deck;

(iii) The towers can be treated as rigid.

Then generalize.

7.4 A fixed concrete platform for offshore oil production rests on the seabed in 200 m of water. An engineer wishes to determine the forces on the platform created by 25 m storm waves, and proposes to model it in a wave tank 1.5 m deep. Find the similarity conditions that the model must obey.

7.5 Fixed platforms cannot be used in very deep water, and are replaced by floating platforms. One kind of floating platform is a SPAR, a cylindrical hull that floats with the cylinder axis vertical, and is moored to the seabed by long cables to anchor piles. An engineer wishes to find the SPAR's response to waves, and is going to test a model in a wave tank. First consider the SPAR as free-floating, so that the mooring system has no significant effect on the way the SPAR responds to waves. Find the similarity conditions that a model must obey.

7.6 Now consider the SPAR as moored. Next the engineer has to determine the forces in the mooring cables induced by storm waves. Find the similarity conditions that a model must obey.

7.7 A SPAR floating platform (Problem 7.5) is constructed in two parts. The hull is carried out on a ship, launched into the water, up-ended, and attached to mooring cables. The topsides are constructed separately, carried out on two barges, and floated over the hull. Some of the ballast water in the hull is then pumped out, so that the hull rises under the topsides and lifts them off the barge. During this operation waves are moving both the barges and the hull, and the relative motion between them determines the impact loads. Find the similarity conditions that a model must obey.

References

1. Baker, JF. Personal communication.

2. Svendsen, IA (1985). Physical modelling of water waves. In *Physical Modelling in Coastal Engineering*, pp. 13–47. Rotterdam: AA Balkema.

3. Heyman, J. Personal communication.

4. Heyman, J (1998). *Structural Analysis: A Historical Approach.* Cambridge: Cambridge University Press.

5. Fitchen, J (1961). *The Construction of Gothic Cathedrals*. Oxford: Clarendon Press.

6. Waldrain, PJ (1935). Science and architecture: Wren and Hooke. *Journal of the Royal Institute of British Architects*, 42, 558.

7. Fairbairn, W (1849). *An Account of the Construction of the Britannia and Conway Tubular Bridges*. London: Longmans.

8. Hossdorf, H (1974). *Model Analysis of Structures*, C Van Amerongen (trans.). New York: Van Nostrand Reinhold.

9. Pippard, AJS and JF Baker (1957). *The Analysis of Engineering Structures*. London: Edward Arnold.

10. Palmer, AC (1976). *Structural Mechanics*. Oxford: Oxford University Press.

11. Palmer, AC, G Hutchinson and JE Ells (1974). Configuration of submarine pipelines during laying operations. *American Society of Mechanical Engineers, Journal of Engineering for Industry*, 96, 1112–1118.

12. Kaustinen, OM, RJ Brown and AC Palmer (1983). Submarine pipeline crossing of M'Clure Strait. *Proceedings of the Seventh International Conference on Port and Ocean Engineering under Arctic Conditions*, Helsinki, VTT Espoo, Vol. 1, pp. 289–299.

13. Brown, RJ and AC Palmer (2007). Developing innovative deep water pipeline construction techniques with physical models. *American Society of Mechanical Engineers, Journal of Offshore Mechanics and Arctic Engineering*, 129, 56.

14. Wenham, F (1866). Aerial locomotion. First Annual Report of Aeronautical Society.

15. Bradshaw, P (1970). *Experimental Fluid Mechanics*. Oxford: Pergamon.

16. Batchelor, GK (1967). *An Introduction to Fluid Dynamics*. Cambridge: Cambridge University Press.

17. Hoerner, SF (1958). *Fluid Dynamic Drag*, published by the author.

18. Urick, RJ (1983). *Principles of Underwater Sound*. New York: McGraw-Hill.

19. European Transonic Windtunnel (2007). www.etw.de.

20. Babinsky, H. Personal communication.

21. McDowell, DM and BA O'Connor (1977). *Hydraulic Behaviour of Estuaries*. London: Macmillan.

22. Palmer, AC and JR Rice (1973). The growth of slip surfaces in the progressive failure of overconsolidated clay. *Proceedings of the Royal Society, Series A*, 332, 527–548.

23. Palmer, AC (1991). Centrifuge modelling of ice and brittle materials. *Canadian Geotechnical Journal*, 28, 896–898.

24. Palmer, AC, DJ White, AJ Baumgard, MD Bolton, AJ Barefoot, M Finch, T Powell, AS Faranski and JAS Baldry (2003). Uplift resistance of buried submarine pipelines, comparison between centrifuge modelling and full-scale tests. *Geotechnique*, 53, 877–883.

Chapter 8

Solutions to Problems

8.1   Introduction

This chapter gives solutions to some of the problems posed at the end of earlier chapters. Solutions to odd-numbered problems in each chapter are given in full. Solutions of even-numbered problems are not given in full, to leave something over for the curiosity and imagination of the reader, but in some cases there are references to fuller discussion elsewhere.

Be aware that there is not always a single right answer.

Solutions to problems in Chap. n are given in Sec. 8.n. The solution to Problem $n.m$ is Subsec. 8.$n.m$, There are no problems in Chap. 1.

8.2   Solutions to problems in Chapter 2

8.2.1

$$1 \text{ year} = 1 \text{ year} \times \frac{365 \text{ days}}{1 \text{ year}} \times \frac{24 \text{ h}}{1 \text{ day}} \times \frac{3600 \text{ s}}{1 \text{ h}} = 3.15 \times 10^7 \text{ s}$$

$$1 \text{ chain} = 1 \text{ chain} \times \frac{22 \text{ yards}}{1 \text{ chain}} \times \frac{3 \text{ ft}}{1 \text{ yard}} \times \frac{0.3048 \text{ m}}{1 \text{ ft}} = 20.117 \text{ m}$$

$$1 \text{ mile} = 1 \text{ mile} \times \frac{5280 \text{ ft}}{1 \text{ mile}} \times \frac{0.3048 \text{ m}}{1 \text{ ft}} = 1609.344 \text{ m}$$

$$1 \text{ acre} = 1 \text{ acre} \times \frac{4840 \text{ square yards}}{1 \text{ acre}} \times \left(\frac{3 \text{ ft}}{1 \text{ yard}}\right)^2 \times \left(\frac{0.3048 \text{ m}}{1 \text{ ft}}\right)^2$$

$$= 4046.856 \text{ m}^2$$

$$1\text{acre ft} = 1 \text{ acre} \times 1\text{ft} = 4046.856\,\text{m}^2 \times 1\,\text{ft} \times \frac{0.3048\,\text{m}}{1\,\text{ft}} = 1233.482\,\text{m}^3$$

$$1\,\text{ft}^3/\text{min} = \frac{1\,\text{ft}^3}{1\,\text{min}} \times \frac{60\,\text{min}}{1\,\text{h}} \times \frac{(0.3048\,\text{m})^3}{(1\,\text{ft})^3} = 1.70\,\text{m}^3/\text{h}$$

$$1\text{mile/h} = \frac{1\,\text{mile}}{1\text{h}} \times \frac{5280\,\text{ft}}{1\,\text{mile}} \times \frac{0.3048\,\text{m}}{1\,\text{ft}} \times \frac{1\,\text{h}}{3600\,\text{s}} = 0.447\,\text{m/s}$$

$$1\text{ knot} = \frac{1\,\text{nautical mile}}{1\,\text{h}} = \frac{1\,\text{nautical mile}}{1\,\text{h}} \times \frac{1853.2\,\text{m}}{1\,\text{nautical mile}}$$

$$\times \frac{1\,\text{h}}{3600\,\text{s}} = 0.515\,\text{m/s}$$

$$1\text{ UK ton} = 1\text{ ton} \times \frac{2240\,\text{lb}}{1\,\text{ton}} \times \frac{0.45359237\,\text{kg}}{1\,\text{lb}} = 1016.047\,\text{kg}$$

$$1\text{kg/m}^3 = \frac{1\,\text{kg}}{1\,\text{m}^3} \times \frac{1\,\text{lb}}{0.45359237\,\text{kg}} \times \left(\frac{0.3048\,\text{m}}{1\,\text{ft}}\right)^3 = 6.243 \times 10^{-2}\,\text{lb/ft}^3$$

$$1\text{lbf/in.}^2 = \frac{1\,\text{lbf}}{1\,\text{in.}^2} \times \frac{0.45359237\,\text{kgf}}{1\,\text{lbf}} \times \frac{9.80665\,\text{N}}{1\,\text{kgf}} \times \left(\frac{1\,\text{in.}}{0.0254\,\text{m}}\right)^2$$

$$= 6.895 \times 10^3\,\text{N/m}^2 = 6.895 \times 10^3\,\text{Pa}$$

$$1\text{ lbf/in.}^2 = (6.895 \times 10^3\,\text{Pa}) \times \frac{1\,\text{MPa}}{10^6\,\text{Pa}} = 6.895 \times 10^{-3}\,\text{MPa}$$

$$1\text{ mm Hg} = 13590\,\text{kgf/m}^3 \times 1\,\text{mm} \times \frac{9.80665\,\text{N}}{1\,\text{kgf}} \times \frac{1\,\text{m}}{1000\,\text{mm}}$$

$$= 133.3\,\text{N/m}^2 = 133.3\,\text{Pa}$$

$$1\text{ horsepower} = \frac{550\,\text{lbf} \times 1\,\text{ft}}{1\,\text{s}} \times \frac{0.45359237\,\text{kgf}}{1\,\text{lbf}} \times \frac{9.80665\,\text{N}}{1\,\text{kgf}} \times \frac{0.3048\,\text{m}}{1\,\text{ft}}$$

$$= 745.7\,\text{Nm/s} = 745.7\,\text{W}$$

$$1\text{ tonne force} = 1\text{ tonne} \times \frac{2000\,\text{kgf}}{1\,\text{tonne}} \times \frac{9.80665\,\text{N}}{1\,\text{kgf}} = 19613.3\,\text{N}$$

$$1\text{lb/ft} = \frac{1\,\text{lb}}{1\,\text{ft}} \times \frac{0.45359237\,\text{kg}}{1\,\text{lb}} \times \frac{1\,\text{ft}}{0.3048\,\text{m}} = 1.488\,\text{kg/m}$$

8.2.3   Heat flow rate is flow of energy per unit time, and its dimensions are those of power, measured in W (watts) in the SI system. Heat flow rate per unit area has dimension [power/area], and is therefore measured in W/m^2. Temperature gradient has dimension $[\Theta/\text{L}]$, and is measured in

$°C/m$. The unit for thermal conductivity is therefore

$$\frac{W/m^2}{°C/m} = W/m\,°C$$

1 Btu raises the temperature of 0.4536 kg of water by 1 °F. Therefore

$$1\,\text{Btu} = 0.4536\,\text{kg}\,°F \times \frac{1\,°C}{1.8\,°F} \times 4185.5\,\frac{J}{\text{kg}\,°C} = 1054.75\,J$$

$$\frac{1\,\text{Btu}}{1\,\text{h}} = \frac{1054.75\,J}{3600\,\text{s}} = 0.293\,W$$

$$\frac{1\,\text{Btu}}{1\,\text{h}\,1\,\text{ft}^2} = \frac{0.293\,W}{1\,\text{ft}^2} \times \frac{(1\,\text{ft})^2}{(0.3048\,\text{m})^2} = 3.154\,W/m^2$$

$$\frac{1\,°F}{1\,\text{cm}} = \frac{1\,°F}{1\,\text{cm}} \times \frac{1\,°C}{1.8\,°F} \times \frac{100\,\text{cm}}{1\,\text{m}} = 55.56\,°C/m$$

$$\frac{1\,\text{Btu}/\text{h}\,\text{ft}^2}{1\,°F/\text{cm}} = \frac{3.154\,W/m^2}{55.56\,°C/m} = 0.0568\,W/m\,°C$$

8.2.5

$$3.24\,\$/\text{US gallon} = \frac{3.24\$}{1\,\text{US gallon}} \times \frac{1\,\text{UK£}}{2.08\$} \times \frac{1\,\text{US gallon}}{0.8\,\text{UK gallon}}$$

$$\times \frac{1\,\text{UK gallon}}{0.00454609\,\text{m}^3} \times \frac{0.001\,\text{m}^3}{1\,\text{l}}$$

$$= 0.428\,£/\text{l}$$

8.3 Solutions to problems in Chapter 3

Problems on dimensionless groups

8.3.1 q/ND^3

8.3.3 U/ND

8.3.5 Density ρ has dimension $[ML^{-3}]$ and surface tension S has dimension $[MT^{-2}]$. None of the other variables has a dimension that includes $[M]$. It follows that a dimensionless group that includes S must also include ρ.

The ratio S/ρ has dimension $[L^3T^{-2}]$. $S/\rho a U^2$ is a group that includes the surface tension effect but excludes the gravitation effect. Alternatively, a group that excludes the velocity U is $S/\rho g a^2$.

8.3.7

(a) Q has the dimension of power/temperature, $[ML^2T^{-3}\Theta^{-1}]$. V is volume $[L^3]$. W and A are areas $[L^2]$. The formula is not dimensionally consistent.

(b) T is force $[MLT^{-2}]$. Q is power $[ML^2T^{-3}]$. v is velocity $[LT^{-1}]$. Both sides of the equation therefore have the same dimension $[MLT^{-2}]$, and the formula is dimensionally consistent but other factors such as yield stress are also significant.

(c) u is velocity $[LT^{-1}]$. A is area $[L^2]$. S is dimensionless. L is length $[L]$. The formula is not dimensionally consistent.

(d) f is frequency, the number of times something happens in unit time, and its dimension is therefore $[T^{-1}]$. c is velocity $[LT^{-1}]$. S is area $[L^2]$. V is volume. L is length $[L]$. The formula is dimensionally consistent.

8.3.8 Only two of the formulas are dimensionally correct. One of the dimensionally correct formulas is excluded by a symmetry argument: a and b are the two shorter sides of a right-angled triangle, and if we exchange a and b the natural frequency cannot change, and therefore the formula must give the same result.

8.3.9 Important are the airplane's linear dimension, mass (because it determines the weight), and control surface positions. The air density and viscosity are important (because they determine Reynolds number), but compressibility and the specific heat ratio are only significant if the aircraft speed approaches the speed of sound: see Chaps. 7 and 4. The numbers of crew and passengers are important only in so far as they influence the total mass.

8.3.11 All the factors except the last are important. The oil composition is probably unimportant except for its influence on the energy release, but heavy oils might not burn completely and therefore would have an effect on the amount of particulates in the smoke.

8.3.13 The dimensions of the pump and the rate at which the handle is pumped determine the flow rate. If the soil is impermeable, flow into the well may limit the amount that can be pumped at one time, and the soil permeability will be significant: see Problem 4.12. If the water table is some way below the ground surface, cavitation may occur in the tubing, because the total pressure falls near 0.

8.3.15 The dimensions of the dam, the permeability of the earth it is made from, and some geotechnical properties such as cohesion and internal friction will be important.

8.3.17 The principal factors are the dimensions of the split, the pressure in the pipeline, the height profile of the pipeline on either side of the leak, and the distances to the valves at which the flow can be shut off. The viscosity of the oil is a secondary factor, and so too is the bubble point, the pressure at which gas begins to come out of solution in the oil.

8.3.19 The thickness of the paint layer, the chemical composition of the paint, and the velocity of air flowing over the paint. If the surface where the paint was applied is porous, then the permeability of the surface might be of secondary significance.

8.3.21 The temperature of the oven, the temperature of the lamb when it was put into the oven, the linear dimensions of the lamb, the heat capacity per unit volume of the meat, the thermal conductivity (Problem 2.3), and the surface heat transfer coefficient (Sec. 4.6) are all important. In a fan oven hot air flows across the surface of the meat, heat transfer at the surface is made faster, and the meat heats up more quickly.

8.3.23 The density and viscosity of the molten glass and the molten tin, the dimensions of the gate that controls the flow of the glass onto the tin, and the rate at which plate glass is drawn off will be important. The flatness will be controlled by the density and viscosity of the glass, and by the time the glass is held at a high temperature (since the viscosity of glass is highly dependent on temperature).

8.4 Solutions to problems in Chapter 4

8.4.1 The author's measurement in central London are listed in the table and plotted in the graph (Fig. 8.1).

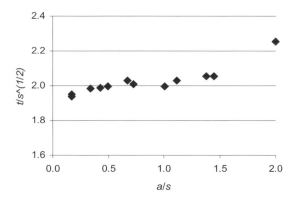

Fig. 8.1. *Pendulum measurements (note suppressed zero on ordinate).*

m	kg	0.91	0.91	0.91	0.91	0.91	0.91	0.91	0.08	0.08	0.08	0.91	0.91
s	m	1.93	1.93	1.94	0.79	0.79	0.79	0.79	2.4	2.01	2.01	6.84	6.84
a	m	1.4	0.66	2.68	0.79	1.57	1.14	0.53	1.03	1	2.24	1.14	1.14
a/s		0.73	0.34	1.38	1.01	2.00	1.45	0.68	0.43	0.50	1.12	0.17	0.17
t	s	2.791	2.758	2.863	1.769	1.997	1.819	1.797	3.075	2.825	2.872	5.099	5.066
$t/\sqrt{s}$	$s/\sqrt{m}$	2.009	1.985	2.056	1.996	2.254	2.053	2.028	1.987	1.995	2.028	1.950	1.937

When a/s is small, $t/\sqrt{s}$ is about $1.99\,\text{s}/\text{m}^{1/2}$. g has been measured at Teddington (a London suburb) and found to be $9.811818\,\text{m}/\text{s}$.[1] Taking this value $t/\sqrt{(s/g)}$ is 6.23. This can be compared with the theoretical value of 2π (6.284) when a/s is small. The two points that correspond to slightly lower values of $t/\sqrt{s}$ are from the longest of the pendulums, and may have been affected by flexibility of the support and the pendulum wire. More refined experiments ought to eliminate that flexibility. The increase of $t/\sqrt{s}$ for larger values of a/s is a real effect, which again comes out of the theory.

8.4.3 Dimensions

$$s \quad [L]$$
$$t \quad [T]$$
$$g \quad [LT^{-2}]$$
$$m \quad [M]$$
$$a \quad [L]$$

Dimension [M] appears only in the dimension of m, and so it is not possible to construct a dimensionless group containing m and any of the other variables. We are left with two dimensions and four variables, so there are two independent groups. [T] appears only in t and g, and so t and g must go together, as the product gt^2, which has dimension [L]. The two dimensionless groups are s/gt^2 and a/s, or any products of powers of those groups. Accordingly

$$\frac{s}{gt^2} = f^{**}\left(\frac{a}{s}\right)$$

where f^{**} is an unknown function, and so

$$s = gt^2 f^{**}\left(\frac{a}{s}\right)$$

and for a fixed value of a/s, s is proportional to t^2 for a fixed value of g, and proportional to g for a fixed value of t. These are the same results as before.

8.4.5   Dimensions

$$
\begin{array}{ll}
q & [L^3T^{-1}] \\
\rho & [ML^{-3}] \\
g & [LT^{-2}] \\
H & [L] \\
\theta & [0]
\end{array}
$$

Dimension [M] appears only in the dimension of ρ and it is therefore not possible to construct a dimensionless group containing ρ and any of the other variables. Look for a group $qH^\alpha g^\beta$. Then

$$[qH^\alpha g^\beta] = [L^3T^{-1}][L]^\alpha[LT^{-2}]^\beta$$

and if the group is dimensionless

$$0 = 3 + \alpha + \beta$$
$$0 = -1 - 2\beta$$

and so α is $-5/2$ and β is $-1/2$. The dimensionless group is $q/(H^{5/2}g^{1/2})$ and therefore

$$\frac{q}{H^{5/2}g^{1/2}} \text{ is a function of } \theta$$

and, for a given notch angle θ, q is proportional to $H^{5/2}$.

8.4.7 Dimensions

$$
\begin{array}{ll}
N & [T^{-1}] \\
\rho & [ML^{-3}] \\
p & [ML^{-1}T^{-2}] \\
R & [L]
\end{array}
$$

The usual argument shows that $N\rho^{1/2}p^{-1/2}R$ is a dimensionless group. It is not possible to make a dimensionless group with only three of the variables. There is no other group for $N\rho^{1/2}p^{-1/2}R$ to be a function of. It follows that $N\rho^{1/2}p^{-1/2}R$ must be a constant, and so

$$
N = \frac{C}{R}\sqrt{\frac{p}{\rho}}
$$

where C is a constant not determined by dimensional analysis. More elaborate theory[2,3] shows the constant to be $2\pi\sqrt{(3\gamma)}$, where γ is the ratio between the specific heat at constant pressure and the specific heat at constant volume.

S has dimensions $[MT^{-2}]$. S/pR is a dimensionless group, though not the only one possible. Accordingly, if surface tension is included

$$
N = \frac{1}{R}\sqrt{\frac{p}{\rho}}f\left(\frac{S}{pR}\right)
$$

where f is an unknown function.[2]

8.4.9 A family of geometrically similar pumps is completely described by the rotor diameter D. Q/ND^3 is one dimensionless group (derived in Problem 3.1). A second group will contain p but will be easier to use if it does not contain Q. $p/\rho N^2 D^2$ is such a group, and a graph of $p/\rho N^2 D^2$ against Q/ND^3 will characterize the whole family of pumps.

Reynolds number is not significant for pumps, unless the fluid pumped is very viscous. If cavitation within the pump is possible, the absolute pressure at the suction side may be important. If the fluid pumped is compressible, the absolute pressures and the pressure ratio are important.

8.4.11   Dimensions

$$
\begin{array}{ll}
E & [\mathrm{ML^2T^{-2}}] \\
r & [\mathrm{L}] \\
t & [\mathrm{T}] \\
\rho & [\mathrm{ML^{-3}}] \\
\gamma & [0]
\end{array}
$$

By inspection, Et^2 has dimension $[\mathrm{ML^2}]$, Et^2/ρ has dimension $[\mathrm{L^5}]$, and $Et^2/\rho r^5$ is dimensionless. It follows that

$$\frac{Et^2}{\rho r^5} \quad \text{is a function of } \gamma$$

and therefore that

$$r = t^{2/5} \sqrt[5]{\frac{E}{\rho}} f(\gamma)$$

where f is an unknown function. The table below lists values of r scaled from the photographs, and the corresponding values of $r/t^{2/5}$. The graph plots r against t on logarithmic scales (Fig. 8.2).

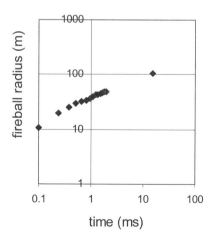

Fig. 8.2. Fireball radius as a function of time.

time t	ms	0.1	0.24	0.38	0.52	0.66	0.8	0.94	1.08
radius r	m	10.9	20.1	25.7	29.7	32.2	34.2	37.2	39.9
$r/t^{(2/5)}$		27.3	35.6	37.8	38.6	38	37.4	38.2	38.7

time t	ms	1.22	1.36	1.5	1.65	1.79	1.93	15
radius r	m	42.3	43.5	45.3	46.5	48.1	49.5	108
$r/t^{2/5}$		39	38.4	38.5	38	38.1	38.1	36.5

The measured diameter is the horizontal diameter across the broadest part of the fireball. The bomb was exploded at the top of a steel tower 30 m high, and after 0.5 ms the fireball reached the ground and was no longer spherical.

It can be seen that r is linearly proportional to $t^{2/5}$. Taking $r/t^{2/5}$ as $38\,\text{m}/(\text{ms})^{2/5}$ and guessing f as 1,

$$\frac{r^5}{t^2} = (38)^5 \frac{\text{m}^5}{(\text{ms})^2} \times \left(\frac{1000\,\text{ms}}{1\,\text{s}}\right)^2 = 7.92 \times 10^{13}\,\text{m}^5/\text{s}^2$$

and so

$$E = \frac{\rho r^5}{t^2} = 7.92 \times 10^{13}\,\text{m}^5/\text{s}^2 \times 1.25\,\text{kg}/\text{m}^3 = 9.9 \times 10^{13}\,\text{J}$$

The energy of atomic bombs is customarily measured as an equivalent in tonnes of the explosive TNT (trinitrotoluene), obviously not an SI unit. One tonne of TNT corresponds to 4.25×10^9 J. The calculated energy is therefore 23,300 tonnes of TNT.

It needs to be emphasized that assigning the value 1 to f is no more than a guess, and that guesses like that are not to be relied on. Taylor's complete calculation[4] found f to be 0.856 for γ equal to 1.4, the value for air. His calculated value was 16,800 tonnes of TNT, consistent with the approximate value of 20,000 tonnes announced by President Truman.

The mass of the bomb itself was about 2000 kg.[5] A sphere of $1.25\,\text{kg}/\text{m}^3$ air that has the same mass as the bomb has a radius of 7.3 m. Except at the very beginning, the mass of air in the fireball is very much greater than the mass of the original bomb.

Atomic bombs plainly raise ethical questions that are far more important than any dimensional analysis.

8.4.13

(i) The dimensions of the variables are

J	$[ET^{-1}]$
s	$[L]$
$\theta_0 - \theta_f$	$[\Theta]$
ρ	$[ML^{-3}]$
U	$[LT^{-1}]$
c	$[EM^{-1}\Theta^{-1}]$
μ	$[ML^{-1}T^{-1}]$
k	$[EL^{-1}T^{-1}\Theta^{-1}]$

(ii) There are eight variables and five fundamental dimensions, and we therefore expect three independent groups.

(iii) Section 4.6 was concerned with heat transfer from a pipeline, and derived a dimensionless group $q/k(\theta_0 - \theta_s)$ for the heat flow rate per unit length q, whose dimensions were $[EL^{-1}T^{-1}]$. Here we are concerned with the total heat flow rate J from a body characterized by a linear dimension s, and so J/s has the same dimension as q and $q/sk(\theta_0-\theta_s)$ is the corresponding dimensionless group. There is a choice of groups for the other variables, but we found earlier that Reynolds number $\rho Us/\mu$ was useful, and so we use that again. Four of the variables, ρ, c, μ, and k, refer to different properties of the fluid. Looking for a dimensionless group that includes them, c and k must appear as the ratio c/k, since otherwise the group is not dimensionless in $[E]$ or $[\Theta]$. The dimension of c/k is $[M^{-1}LT]$, whereas the dimension of μ is $[ML^{-1}T^{-1}]$, and so $c\mu/k$ is another dimensionless group, called Prandtl number Pr, after Ludwig Prandtl (1875–1953), a German fluid dynamicist, who was the first to identify and develop the idea of boundary layers. (A nice story about Prandtl is that he was somewhat naive, and when he was 34 he thought he ought to get married. He went to his professor and asked to marry his professor's daughter. He neglected to say which daughter he had in mind, and so the professor and his wife decided which of their daughters he should marry. They were very happy.)

Prandtl number has a physical interpretation. The diffusion of heat is characterized by a thermal diffusivity $k/\rho c$, dimension $[L^2 T^{-1}]$. The diffusion of the shear stress that determines the distribution of velocity near a solid surface is characterized by a kinematic viscosity μ/ρ, again with dimension $[L^2 T^{-1}]$. Pr is therefore the ratio between the diffusivity of heat and the diffusivity of velocity, loosely, the ratio between the rate at which velocity diffuses and the rate at which temperature diffuses. In many liquid and gases Pr is close to 1, and then near the surface of a heated body in a free stream the velocity boundary layer and the thermal boundary layer are similar; and that similarity (Reynolds' analogy) can be exploited in analysis and experiments. In liquid metals like mercury and molten sodium, on the other hand, Pr is very much smaller, around 0.01 (because the viscosity is low and the thermal conductivity high), and there the temperature boundary layer is much thicker than the velocity boundary layer. The reverse is true for viscous lubricating oils.

(vi) Putting this together, we conclude that

$$\frac{q}{sk(\theta_0 - \theta_f)} = f\left(\frac{\rho U s}{\mu}, \frac{c\mu}{k}\right)$$

or since the first group is a Nusselt number Nu

$$Nu = f(Re, Pr)$$

Many experimental and theoretical results in heat transfer are economically expressed in this way.[6,7] There are other useful groups, among them Péclet number, the product RePr. Heat transfer has an extensive literature: see, for example, Prandtl,[6] Richardson,[8] Rogers and Mayhew,[7] and many more recent books.

(v) If kinetic energy can be transformed into heat, [E] is not an independent dimension but instead $[ML^2 T^{-2}]$, and so now we have four dimensions and eight variables, and therefore four dimensionless groups rather than three. Again there are choices, but a convenient group is Eckert number $U^2/c\theta$, after Ernst Eckert (1904–2004), where θ is a temperature. If the flow is in a perfect gas, and θ is the absolute temperature, thermodynamics shows that Eckert number is the square of the Mach number Ma introduced earlier, multiplied by $\gamma - 1$, where γ is the specific heat at constant pressure divided by the specific heat at constant volume.

8.4.15   Dimensions

$$N \quad [T^{-1}]$$
$$L \quad [L]$$
$$m \quad [ML^{-1}]$$
$$F \quad [ML^3[T^{-2}]$$

$N^2L^4mF^{-1}$ is dimensionless, and since there is nothing else for it to be a function of it must be a constant, but there will be different constants for different modes.

$$N = \frac{C_n}{L^2}\sqrt{\frac{m}{F}}$$

where C_n has a different value for each mode. A full analysis can be found in standard books on vibration dynamics.

8.5   Solutions to problems in Chapter 5

8.5.1   The variables and their dimensions are

propagation pressure p	$[FL^{-2}]$
pipe diameter D	$[L]$
wall thickness t	$[L]$
yield stress Y	$[FL^{-2}]$

taking force [F] as a fundamental dimension. The dimensionless groups are p/Y and D/t. The graph below plots p/Y against D/t on logarithmic scales (Fig. 8.3). If force is given the dimension $[MLT^{-2}]$ the groups are the same.

In the example given, D/t is $660.4/19.05 = 34.7$. Reading from the graph, $p/Y = 0.0042$, and so p is $0.0042 \times 483\,\text{MPa} = 2.029\,\text{MPa}$. Seawater has a density of about $1025\,\text{kg/m}^3$, and therefore at a depth d m the pressure is $1025\,\text{kg/m}^3 \times d \times 9.81\,\text{m/s}^2 = 10055\,d\,\text{Pa} = 0.01055\,d\,\text{MPa}$, and therefore the water depth corresponding to the propagation pressure is $2.029/0.01055 = 200\,\text{m}$.

Elastic modulus E has dimension $[FL^{-2}]$, and a third dimensionless group is simply E/Y.

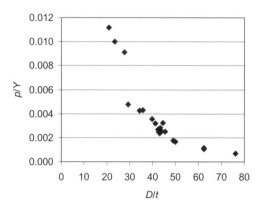

Fig. 8.3. Propagation pressure data.

8.6 Solutions to problems in Chapter 6

8.6.1   Call the displacement of the left-hand mass X_1, measured from the equilibrium position, and the displacement of the right-hand mass X_2. The tension in the left-hand spring is then kX_1, the tension in the center spring is $k(X_2 - X_1)$, and the tension in the right-hand spring is $-kX_2$. The equations of motion for the two masses are

$$m\frac{d^2X_1}{dT^2} = k(X_2 - X_1) - kX_1 \tag{8.6.1}$$

$$m\frac{d^2X_2}{dT^2} = -k(X_2 - X_1) - kX_2 \tag{8.6.2}$$

As before, nondimensionalize T with respect to $\sqrt{m/k}$, and X_1 and X_2 with respect to an arbitrary length a. Then (8.6.1) and (8.6.2) become

$$\frac{d^2x_1}{dt^2} = -2x_1 + x_2 \tag{8.6.3}$$

$$\frac{d^2x_2}{dt^2} = x_1 - 2x_2 \tag{8.6.4}$$

The natural frequencies correspond to solutions of the form

$$x_1 = a\sin\omega t \tag{8.6.5}$$

$$x_2 = \lambda a\sin\omega t \tag{8.6.6}$$

Substituting in (8.6.5) and (8.6.6)

$$-\omega^2 = -2 + \lambda \qquad (8.6.7)$$

$$-\lambda\omega^2 = 1 - 2\lambda \qquad (8.6.8)$$

Eliminating ω^2 by multiplying the first equation by λ and subtracting the second

$$0 = \lambda^2 - 1 \qquad (8.6.9)$$

which has two roots, $+1$ and -1. The first root corresponds to a mode in which the two masses move in phase, in the same direction and with the same amplitude. Substituting back into (8.6.7), the corresponding value of ω^2 is 1. The second root corresponds to a mode in which the two masses move out of phase, with the same amplitude but in opposite directions, so that one is moving to the left while the other is moving to the right. The corresponding value of ω^2 is 3. The corresponding dimensionless periods are 2π and $2\pi/\sqrt{3}$, and the periods can be found by multiplying by $\sqrt{m/k}$.

8.6.3

(i) E has dimensions $[FL^{-2}]$, and so D has dimensions $[FL]$. ρg is force per unit volume, and its dimensions are $[FL^{-3}]$. It follows that $D/\rho g$ has dimension $[L^4]$, and therefore that $\sqrt[4]{\frac{D}{\rho g}}$ is a characteristic length, denoted λ. That suggests that we could try nondimensionalizing the lengths x, y, and w, so that

$$x' = \frac{x}{\lambda} \qquad (8.6.10)$$

and y' and w' are defined similarly. In the new variables the governing equation becomes simply

$$\frac{\partial^4 w'}{\partial x'^4} + 2\frac{\partial^4 w'}{\partial x'^2 \partial y'^2} + \frac{\partial^4 w'}{\partial y'^4} = w' \qquad (8.6.11)$$

which does not contain D or ρg explicitly.

(ii)

$$\varepsilon_{xx} = (t/2)\frac{\partial^2 w}{\partial x^2} = \frac{t}{2\lambda}\frac{\partial^2 w'}{\partial x'^2} \qquad (8.6.12)$$

A point load P is balanced by increased buoyancy induced by downward deflection of the ice. From equilibrium

$$P = \int \rho g w \, dx \, dy = \rho g \lambda^3 \int w' \, dx' \, dy' \qquad (8.6.13)$$

where the integrals are taken over the whole area that deflects. A point load does not bring any other length into the problem, and therefore the relationship between w', x', and y' is universal, and does not depend on the particular value of λ. From (8.6.13), $\int w' \, dx' \, dy'$ is proportional to $P/\rho g \lambda^3$, and so the largest value of $\frac{\partial^2 w'}{\partial x'^2}$ must also be proportional to $P/\rho g \lambda^3$. From (8.6.12), the largest value of the strain is

$$\max(\varepsilon_{xx}) = \frac{t}{2\lambda} \max\left(\frac{\partial^2 w'}{\partial x'^2}\right) \propto \frac{t}{\lambda} \frac{P}{\rho g \lambda^3} \qquad (8.6.14)$$

and is therefore proportional to P/Et^2. It has not been necessary to solve the governing equation.

(iii) The model predicts that the maximum strain is proportional to the load P and inversely proportional to the square of the ice thickness t. It follows that if failure occurs when a critical strain is reached, the critical value of the load is proportional to the thickness squared, which is consistent with the rule.

(iv) The characteristic length $\sqrt[4]{Et^3/(12(1 - \nu^2)\rho g)}$ governs the horizontal extent of vertical deflections. For ice, E is about 8 GN/m^2 and ν is about 0.3. For seawater, ρg is about 10 kN/m^3. If the ice is 1 m thick, the characteristic length is 16 m. If the ice is t m thick, the characteristic length is $16t^{3/4}$ m. We would expect that spreading out a load would not have much effect unless the distance over which it is spread out is comparable with the characteristic length.

The empirical formula in (ii) suggests that a person weighing 70 kg (say) might break through the ice when it is less than 45 mm thick. The corresponding characteristic length is 1.5 m. That is consistent with the advice that if you are trying to save somebody who has fallen through ice, you ought to spread out your weight and lie flat on the ice and crawl.

8.7   Solutions to problems in Chapter 7

8.7.1   If the anchor is moving slowly, and the test is to be conducted under water, the factors that will determine how far the anchor will dig into the rock are the submerged weight W of the anchor, the submerged weight ρg per unit volume of the rock (where ρ is mass per unit volume and g is gravitational acceleration), and the linear dimensions ℓ of the anchor and the rock fragments. A scale model ought therefore to have the dimensionless group $W/\rho g \ell^3$ the same in the model as in the prototype. Under water, this can be achieved by making the model anchor of the same metal as the prototype anchor, and the model rock of the same rock as the prototype. The size of the rock fragments should be scaled in the same way as the linear dimensions of the anchor.

In air, it is still possible to maintain $W/\rho g \ell^3$ similarity, but either the material density of the rock ρ_{rock} or the material density of the anchor ρ_m must be changed. That is because the rock is much less dense than water, density ρ_w, and so in air the weight of the anchor is increased in the ratio $\rho_m/(\rho_m - \rho_w)$ compared to under water, whereas the weight of the rock is increased in the larger ratio $\rho_{rock}/(\rho_{rock} - \rho_w)$.

If dynamic effects are important, we also want to maintain Froude similarity, so that Froude number $U/\sqrt{(g\ell)}$ (where U is the velocity of the anchor; see Secs. 3.4 and 4.5) is the same in the model as in the prototype. Froude number represents the ratio between the effects of weight and inertia. In addition, geotechnics tells us that when a particulate material like broken rock deforms the particles tend to move further apart, which means that water has to flow from outside into the increased pore space between the particles. That requires a pressure gradient in the broken rock, and if the rock is impermeable the effect may be to reduce the pressure in the pore water, which increases the force between the particles and modifies the mechanical behavior of the rock pile. The effect is important in cutting and plowing of fine-grained soils at high speeds. Palmer[9] gives an example, and shows that the relevant dimensionless group is $U\ell/\kappa$, where κ is the diffusivity of water in the soil, in geotechnics called the coefficient of consolidation. It is not possible to have both $U/\sqrt{(g\ell)}$ similarity and $U\ell/\kappa$ similarity without changing the fluid. In reality the diffusivity in broken rock is so large that pore pressure effects are insignificant, though they would be important if the pipeline were covered with fine sand or silt.

8.7.3 The flexibility of a long structure allows energy to be transmitted from the wind into structural oscillations. There is more than one kind of oscillation, and mechanisms include vortex-excited oscillation, flutter, galloping and divergence. A suspension bridge derives part of its stiffness from the weight of the deck and part from the structural flexural rigidity of the deck: the first component is present because even if the deck were infinitely flexible a force would still be required to deflect it. For a given geometry, the first component is proportional to mg and the second component to EI/ℓ^3. A model therefore ought to keep the dimensionless ratio $EI/mg\ell^3$ the same as in the prototype.

Vortex-excited oscillations depend on the reduced velocity U/ND, where U is the wind velocity, N is the natural frequency (or frequencies), and D a characteristic transverse dimension of the deck. For a beam whose effective stiffness is the flexural rigidity EI, the natural frequencies N are proportional to $1/\ell^2\sqrt{EI/m}$ (Problem 4.15). Therefore, if D is scaled in the same way as ℓ, $U\ell\sqrt{m/EI}$ should be the same in the model as in the prototype.

The mass of the air around the deck may be significant, and if D is scaled with ℓ then $m/\rho\ell^2$ should be the same, where ρ is the air density. Many modes of oscillation include torsion as well as bending, and this is seen dramatically in the film of the Tacoma Narrows failure. Similarity for torsional oscillations can be secured by keeping EI/GJ constant.

8.7.5 Following from the analysis in Sec. 4.5, the motion will be driven by the waves. We would like Froude number $U/\sqrt{(gL)}$ to be the same in the model as in the prototype, where U is a representative velocity, g the gravitational acceleration, and L is a representative length. The velocity U induced by a wave depends on g, the wave height H, and the wave period t. The four variables and their dimensions are

$$U \quad [LT^{-1}]$$
$$H \quad [L]$$
$$t \quad [T]$$
$$g \quad [LT^{-2}]$$

There are four variables and two dimensions, and therefore two independent dimensionless groups. Taking the two groups as U/gt and H/gt^2, U/gt is a function of H/gt^2, and so

$$U = gtf\left(\frac{H}{gt^2}\right)$$

where f is an unknown function, and

$$\frac{U}{\sqrt{gL}} = \frac{t\sqrt{g}}{\sqrt{L}}f\left(\frac{H}{gt^2}\right)$$

To have the Froude number the same in the model as in the prototype, and since both will have the same value of g, we must keep $t/\sqrt{L}$ the same. If the model is $1/50$ scale, and the prototype waves have height 12 m and period 15 s, the waves in the model wave tank have period $15/\sqrt{50}$ s, which is 2.12 s. The wave height H then scales in the same way as other lengths, which makes $f(H/gt^2)$ the same, and then the model waves are 0.24 m high. If the sea state is represented by wave spectrum rather than by a single regular wave train, the heights and periods of the model test spectrum scale in the same way as the height and period of a single wave train.

Mass scales as L^3, and then the model floats at the correct scaled depth. Dynamic pressures are proportional to U^2 and therefore to L. Areas are proportional to L^2, and so dynamic forces are proportional to L^3 and the correct relationship between mass and dynamic forces is maintained. Accelerations are the same in the model as in the prototype.

For the same reasons as for ship model testing in towing tanks, it is not possible to maintain the same Reynolds number in the model as in the prototype. Surface tension has no significant effect.

8.7.7 The motions of the barges and the hull obey the same conditions as in Problem 7.5, and the modeling principle is the same.

The impact loads are highly localized, and depend on the details of the structure close to the points at which contact is made. It is probably best not to try to analyze the local effects with the same models as the motions of the complete vessels. Instead the impact velocities can be determined from the first models, and then the local effects either by full-scale tests or by model tests on parts of the structure to a larger scale. Structural experiments on impact are discussed by Johnson[10] and Jones.[11]

References

1. Kaye, GWC and TH Laby (1971). *Tables of Physical and Chemical Constants*. London: Longman.

2. Dowling, AP and JE Ffowcs-Williams (1983). *Sound and Sources of Sound*. England, UK: Ellis Horwood.

3. Urick, RJ (1983). *Principles of Underwater Sound*. New York: McGraw-Hill.

4. Taylor, GI (1950). The formation of a blast wave by a very intense explosion. II. The atomic explosion of 1945. *Proceedings of the Royal Society, Series A*, 101, 175–186.

5. Rhodes, R (1986). *The Making of the Atomic Bomb*. New York: Simon & Schuster.

6. Prandtl, L (1952). *The Essentials of Fluid Dynamics*, Blackie, London. Originally published as *Abriss Der Strömungslehre*, Vieweg, Braunschweig (1931).

7. Rogers, GFC and YR Mayhew (1980). *Engineering Thermodynamics, Work and Heat Transfer*. London: Longman.

8. Richardson, LF (1920). Convective cooling and the theory of dimensions. *Proceedings of the Physical Society of London*, 32, 405–409.

9. Palmer, AC (1999). Speed effects in cutting and ploughing. *Geotechnique*, 49(3), 285–294.

10. Johnson, W (1972). *Impact Strength of Materials*. London: Edward Arnold.

11. Jones, N and T Wierzbicki (1983). *Structural Crashworthiness*. St. Louis, MO: Science and Technology Books.